MotoGP: The

Complete

Championship History

The Definitive Guide to 75 Years of Grand Prix Motorcycle Racing History and Heroes—From 1949 to Today

By: **Daryl Whiting**

Dedication

To the brave souls who dance with danger at 200 miles per hour, pushing the boundaries of human capability and mechanical perfection. To the engineers who dream in carbon fiber and titanium, the mechanics whose hands shape victory, and the fans whose passion fuels this magnificent obsession.

This book is dedicated to every rider who has thrown a leg over a Grand Prix motorcycle, knowing that glory and disaster are separated by the width of a tire contact patch. From the pioneers who built this sport with courage and vision, to the modern gladiators who continue to write its future in turns of breathtaking beauty and heart-stopping drama.

Most especially, to those who gave everything to the sport they loved, and whose memory reminds us that behind every lap time is a human story of ambition, sacrifice, and the relentless pursuit of that perfect lap.

"The bike is nothing without the rider, and the rider is nothing without the bike. Together, they become something greater than the sum of their parts."
— **Valentino Rossi**

Table Of Contents

Introduction

The Cathedral of Speed

In the predawn darkness of a race day morning, when the paddock lies silent and the grandstands stand empty, there exists a moment of perfect tranquility at a Grand Prix circuit. The asphalt stretches like a ribbon of black silk, unmarked by tire rubber, unblemished by the drama that will soon unfold. It is in these quiet moments that one can truly appreciate what MotoGP represents: the ultimate expression of human ambition meeting mechanical perfection, played out at speeds that defy both logic and survival instinct.

MotoGP is more than a motorcycle racing championship. It is a theater of dreams where physics meets philosophy, where split-second decisions carry the weight of careers, and where the difference between triumph and tragedy can be measured in millimeters and milliseconds. For over seven decades, this sport has captured the imagination of millions, creating heroes and legends while pushing the very boundaries of what is possible on two wheels.

This is the story of that evolution – from the modest beginnings of the world motorcycle championship in 1949 to the global spectacle that is modern MotoGP. It is a tale written in brake dust and tire smoke, told through the roar of engines and the silence of concentration, illustrated by moments of sublime skill and heart-stopping courage.

The men and women who have shaped this sport are cut from a different cloth. They possess an almost supernatural ability to process information at superhuman speeds, to make calculations that would challenge a computer while leaning at angles that would seem to mock gravity itself. They are athletes, artists, and gladiators rolled into one, wielding 300-horsepower missiles with the precision of a surgeon and the fearlessness of a fighter pilot.

But MotoGP's story is not merely one of individual heroism. It is equally a narrative of technological advancement, of engineering minds pushing the boundaries of materials science, aerodynamics, and electronics. Every component on a MotoGP machine represents thousands of hours of development, computer modeling, and real-world testing. These motorcycles are not merely fast; they are masterpieces of human ingenuity, each one a rolling laboratory that advances the entire motorcycle industry.

The circuits themselves are characters in this story – from the sweeping curves of Mugello that separate the truly gifted from the merely talented, to the night-time drama of Qatar where the season traditionally begins under artificial suns. Each venue presents unique challenges, demanding different skills and rewarding different approaches to the art of motorcycle racing.

Remarkably, MotoGP has evolved from a predominantly European affair to a truly global championship, creating fans and fostering talent on every continent. The sport has become a diplomatic passport, bringing together nations through a shared appreciation for speed, skill, and spectacle. It has transcended cultural boundaries while creating new ones – the brotherhood and sisterhood of those who understand that there is no sensation quite like watching a master at work, threading a motorcycle through a sequence of corners with the kind of precision that makes time seem to slow down.

The economic impact of MotoGP extends far beyond the championship itself. The technologies developed for Grand Prix racing inevitably find their way to road motorcycles, improving safety, performance, and efficiency for millions of everyday riders. The sport serves as a catalyst for innovation, driving advances in materials science, electronics, and aerodynamics that benefit not just the motorcycle industry, but automotive and aerospace applications as well.

Yet for all its technological sophistication and global reach, MotoGP remains fundamentally human. It is about individuals making split-second decisions at the very edge of physical possibility, about teams working in perfect harmony to find that extra tenth of a second, about fans whose passion transcends language and nationality. It is about the moment when a rider finds that perfect line through a corner, when machine and human become one entity in pursuit of the fastest possible lap time.

The championship has weathered many storms – economic downturns, technical controversies, tragic losses, and global pandemics. Through each challenge, the sport has emerged stronger, more refined, and more compelling. The constant evolution of rules and regulations reflects not just a desire for fair competition, but a commitment to safety and sustainability that ensures MotoGP will continue to thrill future generations.

This book chronicles that journey from the first tentative steps of the 1949 championship to the high-technology spectacle of today. It celebrates the pioneers who built the foundation, the innovators who pushed the boundaries, and the athletes who continue to redefine what is possible on a motorcycle. It examines the circuits that have become sacred ground for racing enthusiasts, the technologies that have revolutionized not just racing but motorcycling itself, and the business that has grown around this most demanding of sports.

Most importantly, it attempts to capture the essence of what makes MotoGP so compelling – that unique combination of individual skill, team effort, technological advancement, and pure human drama that unfolds eighteen times a year at circuits around the world. It is a story of passion and precision, of risk and reward, of the eternal human quest to go faster, brake later, and lean further than anyone has before.

As we embark on this journey through MotoGP's rich history, we invite you to experience the sport through the eyes of those who have lived it, shaped it, and been shaped by it in return. From the oil-stained workshops where dreams are built to the victory podiums where legends are crowned, from the technical meetings where championships are won and lost to the millisecond moments where heroes are made, this is the complete story of the world's premier motorcycle racing championship.

Welcome to MotoGP. Welcome to the cathedral of speed.

ONE

The Birth of Grand Prix Motorcycle Racing (1949-1959)

The year 1949 marked a pivotal moment in motorsport history, though few could have predicted the global phenomenon that would emerge from a modest gathering of European motorcycle racing officials. The devastation of World War II had left the sporting world fragmented, with national championships operating in isolation and manufacturers struggling to rebuild their racing programs. It was against this backdrop that the Fédération Internationale de Motocyclisme (FIM) took the bold step of establishing the world's first truly international motorcycle racing championship.

The FIM had been founded in 1904, but the war years had effectively suspended international cooperation in motorcycle sport. As Europe began to rebuild, visionary officials within the organization recognized an unprecedented opportunity. The existing patchwork of national championships – the Isle of Man TT, the French Grand Prix, the Ulster Grand Prix – had proven immensely popular, but lacked cohesion. Manufacturers were investing heavily in racing as a means of demonstrating their engineering prowess and attracting customers to their road bikes, yet their efforts were scattered across dozens of disconnected events.

The proposal was elegantly simple yet revolutionary in its scope: create a unified world championship that would crown annual champions across multiple engine displacement categories, providing a clear hierarchy of achievement while accommodating the diverse range of motorcycles being produced. The inaugural season would feature four classes – 500cc, 350cc, 250cc, and 125cc – each representing different approaches to motorcycle design and appealing to various manufacturer specialties.

The 500cc class was positioned as the premier category from the outset, attracting the most powerful machines and, consequently, the most talented riders. These were the motorcycles that pushed the absolute boundaries of speed and handling, requiring not only engineering excellence but exceptional bravery and skill to master. The 350cc class offered a middle ground, often producing some of the most competitive racing as manufacturers and riders found the sweet spot between power and manageability. The smaller classes, 250cc and 125cc, served as proving grounds for emerging talent while showcasing the precision engineering required to extract maximum performance from limited displacement.

The economic rationale was equally compelling. Manufacturers could demonstrate their technical superiority on an international stage, with victories translating directly into sales prestige and marketing opportunities. For riders, the world championship offered unprecedented career opportunities, with the possibility of competing across multiple continents and establishing truly international reputations.

The 1949 season commenced with six rounds across five countries: Great Britain (Isle of Man TT), Switzerland, Netherlands, Belgium, and Italy. Each venue brought its own character and challenges, from the fearsome 37.7-mile Isle of Man course with its stone walls and mountain sections to the high-speed banked sections of Monza. This diversity would become a hallmark of Grand Prix racing, testing both man and machine across a comprehensive range of conditions.

The championship structure was elegantly simple: points were awarded to the top six finishers in each race, with 8 points for victory, 6 for second, 4 for third, 3 for fourth, 2 for fifth, and 1 for sixth. Only a rider's best results would count toward the championship, a system designed to encourage participation across the full season while accounting for the mechanical unreliability that was commonplace in the era.

***

The inaugural 1949 season produced a champion who embodied the spirit of the new world championship: Leslie Graham, a British rider whose victory aboard his AJS Porcupine 500cc machine established him as the first official Grand Prix world champion. Graham's achievement was particularly significant given the technical challenges he overcame. The AJS Porcupine was an advanced but temperamental machine, featuring a complex twin-cylinder engine with dual overhead camshafts and a unique "Porcupine" cylinder head design that gave the bike its nickname.

Graham's championship campaign demonstrated the qualities that would become essential for Grand Prix success: technical knowledge, adaptability, and unwavering determination. Unlike modern riders who rely on extensive support teams, Graham often worked on his own machine, understanding every component and making critical setup decisions. His victory at the Swiss Grand Prix showcased these skills perfectly, as he overcame early mechanical problems to deliver a masterful ride in challenging mountain weather conditions.

The 1949 season also introduced the world to several other riders who would define the early championship years. Nello Pagani, the Italian veteran, claimed victory at the Nations Grand Prix at Monza, demonstrating the passionate Italian approach to motorcycle racing that would become legendary. Bruno Ruffo took the 250cc championship, establishing the first of what would become many Italian dominations in the smaller classes. These early champions established patterns and traditions that persist in modern MotoGP: the importance of home crowd support, the psychological pressure of defending championship leads, and the crucial role of mechanical reliability in championship campaigns.

The 1950 season saw the emergence of a rider who would dominate the early championship years and set standards that remain impressive today: Geoff Duke. The British rider's approach to motorcycle racing was methodical and professional, contrasting sharply with the more instinctive styles of many contemporaries. Duke brought a scientific approach to bike setup and race strategy that was ahead of its time, working closely with Norton's engineers to develop machines that were not just fast but reliable and predictable.

Duke's first championship in 1950 aboard the Norton Manx demonstrated the importance of manufacturer support in Grand Prix racing. Norton had invested heavily in racing development, understanding that success on the track translated directly to sales success. The Manx was a sophisticated machine featuring an advanced single-cylinder engine with dual overhead camshafts and a revolutionary featherbed frame designed by the McCandless brothers. This combination of power and handling gave Duke a decisive advantage, allowing him to secure three victories and the championship title.

The early years also witnessed the emergence of rivalries that would captivate fans and drive technical development. The battle between British and Italian manufacturers became particularly intense, with Norton, AJS, and Velocette representing British engineering excellence while Gilera and MV Agusta showcased Italian passion and innovation. These manufacturer rivalries extended beyond mere competition, representing different philosophical approaches to motorcycle design and racing strategy.

These early champions established the template for Grand Prix success that remains relevant today. Physical fitness, technical knowledge, mental strength, and the ability to perform under pressure became the fundamental requirements for championship contention. The riders of the 1950s rode without the safety equipment and medical support that modern riders take for granted, yet they established standards of professionalism and dedication that inspired subsequent generations.

The diversity of the early champions also demonstrated the global appeal of motorcycle racing. While Europeans dominated the early years, riders from around the world were drawn to the championship, laying the foundation for the truly international sport that MotoGP would become. Each champion brought unique perspectives and techniques, contributing to the rapid evolution of riding styles and racing strategies that characterized the formative years.

***

The technological landscape of early Grand Prix racing was characterized by diversity and experimentation as manufacturers explored different engineering approaches in their quest for competitive advantage. The late 1940s and early 1950s represented a fascinating period of transition, with traditional four-stroke designs beginning to give way to innovative two-stroke engines that would eventually revolutionize motorcycle racing.

The initial championship years were dominated by sophisticated four-stroke machines that represented the pinnacle of pre-war development combined with post-war innovation. Norton's Manx series exemplified this approach, featuring advanced single-cylinder engines with dual overhead camshafts, sophisticated valve gear, and meticulously crafted components. The Manx's success stemmed from Norton's comprehensive approach to racing development, combining engine power with the revolutionary featherbed frame that provided unprecedented handling characteristics.

Gilera's four-cylinder machines represented another approach to four-stroke development, showcasing Italian engineering flair and the potential of multi-cylinder configurations. The Gilera four was a complex and expensive machine, featuring superb build quality and innovative engineering solutions. Riders like Geoff Duke found the Gilera's power delivery smooth and predictable, though the machine's complexity often led to reliability issues that could decide championship campaigns.

MV Agusta entered Grand Prix racing with similarly sophisticated four-stroke machines, though their greatest successes would come later in the decade. The Italian manufacturer's approach emphasized precision engineering and attention to detail, creating machines that were as beautiful as they were effective. The company's commitment to racing represented more than mere competition; it was a demonstration of Italian engineering excellence and artistic sensibility applied to motorcycle design.

However, the most significant technical development of this era was the gradual emergence of two-stroke technology as a competitive force in Grand Prix racing. While two-stroke engines had existed for decades, their application to high-performance racing had been limited by various technical challenges. The fundamental advantages of two-stroke design – lighter weight, simpler construction, and higher power-to-displacement ratios – were well understood, but achieving reliability and usable power delivery required significant engineering development.

The German manufacturer DKW had experimented with two-stroke racing machines before the war, achieving some success in smaller displacement categories. Their work demonstrated the potential of two-stroke design while highlighting the challenges that needed to be overcome. Post-war development by various manufacturers, including Italian firms like Parilla and Mondial, began to unlock the performance potential of two-stroke engines in racing applications.

The technical advantages of two-stroke engines became increasingly apparent as development progressed. The elimination of valves, camshafts, and complex valve gear significantly reduced weight and mechanical complexity while allowing higher engine speeds. The power stroke occurring with every crankshaft revolution theoretically doubled power output compared to four-stroke engines of similar displacement. These advantages were particularly pronounced in smaller displacement classes, where weight reduction and high power-to-weight ratios provided decisive competitive benefits.

Manufacturing processes also played a crucial role in early technical development. The post-war period saw rapid advances in materials science, precision manufacturing, and quality control procedures. These improvements enabled manufacturers to build more powerful and reliable engines while reducing weight and improving performance consistency. The development of new aluminum alloys, improved bearing materials, and more precise machining techniques allowed significant performance gains across all displacement categories.

Fuel technology represented another critical area of development. The availability of high-octane racing fuels enabled higher compression ratios and more aggressive ignition timing, directly translating to increased power output. Understanding of carburetion and mixture preparation improved dramatically, with manufacturers developing sophisticated carburetor systems that provided optimal fuel delivery across the engine's operating range.

The evolution of tire technology during this period had profound implications for motorcycle racing performance. Early racing tires were relatively primitive by modern standards, with limited grip and short lifespan. However, gradual improvements in rubber compounds and construction techniques provided riders with increased confidence and allowed them to push their machines harder. The relationship between tire performance and motorcycle setup became increasingly important, establishing principles that remain fundamental to modern racing.

Most importantly, the early championship years established the culture of technical innovation that would drive Grand Prix racing throughout its history. Manufacturers understood that racing success required continuous development and innovation, leading to rapid advances in all aspects of motorcycle design. This competitive pressure created an environment where technical excellence was not just rewarded but essential for survival, establishing the foundation for the technological arms race that would characterize subsequent decades.

The collaborative relationship between riders and engineers also developed during this period, with successful combinations like Geoff Duke and Norton's engineering team demonstrating the importance of effective communication between rider feedback and technical development. This partnership approach would become increasingly sophisticated over time, eventually evolving into the comprehensive technical support systems that characterize modern MotoGP.

TWO

Golden Era Champions (1960-1979)

The emergence of Giacomo Agostini in the early 1960s marked the beginning of what many consider the golden age of Grand Prix motorcycle racing. Born in Brescia, Italy, in 1942, Agostini would go on to achieve a level of dominance that remains unmatched in the sport's history. His record of 15 world championships – eight in the 500cc class and seven in the 350cc class – stands as a testament not only to his extraordinary talent but also to his longevity and adaptability across nearly two decades of competition.

Agostini's rise to prominence coincided with MV Agusta's golden period, creating one of the most successful partnerships in motorsport history. When he joined the Italian manufacturer in 1965, MV Agusta was already established as a racing powerhouse, but Agostini's arrival elevated the team to unprecedented heights. His first championship came in 1966 in the 350cc class, but it was his

500cc debut victory in 1966 that truly announced his arrival as a force to be reckoned with.

The technical sophistication of MV Agusta's machines provided Agostini with the perfect platform to showcase his exceptional abilities. The MV Agusta 500cc four-cylinder was a masterpiece of engineering, featuring advanced metallurgy, precise fuel injection systems, and a level of finish that was unmatched in the paddock. The engine produced approximately 70 horsepower at 12,000 rpm, figures that were extraordinary for the era, while the chassis provided the stability and precision required to harness such power effectively.

Agostini's riding style was characterized by smoothness and precision that seemed almost effortless. Where other riders fought their machines, Agostini appeared to dance with his, finding the perfect balance between aggression and control. His cornering technique was revolutionary for the era, using a more upright riding position that allowed greater precision while maintaining higher speeds through technical sections. This approach required exceptional physical fitness and mental concentration, attributes that Agostini developed through rigorous training programs that were advanced for their time.

The statistical dominance of Agostini's career is staggering. Between 1966 and 1975, he won 122 Grand Prix races, a record that stood for decades. His 1971 season was particularly remarkable, winning all races he entered across both 350cc and 500cc classes. This level of dominance was achieved during an era when mechanical reliability was far from guaranteed, making his consistency even more impressive.

Beyond the numbers, Agostini's impact on motorcycle racing extended to his professionalism and approach to the sport. He was among the first riders to treat Grand Prix racing as a full-time profession, dedicating himself completely to physical fitness, technical development, and race preparation. His relationship with MV Agusta's engineers was collaborative and productive, with his detailed feedback contributing significantly to the continuous development of the machines.

Agostini's rivalry with other great riders of the era, particularly Mike Hailwood, produced some of the most memorable racing in Grand Prix history. Their battles on track were characterized by mutual respect and extraordinary skill levels that pushed both riders to their limits. When Hailwood returned to Grand Prix racing in

1978 aboard a Suzuki, his battles with Agostini aboard the Yamaha created instant classics that remain legendary among fans.

The Italian's decision to leave MV Agusta in 1973 and join Yamaha marked a significant moment in Grand Prix history. The move was motivated by Agostini's desire to compete with two-stroke technology, recognizing that the future of Grand Prix racing lay with these lighter, more powerful machines. His adaptation to two-stroke engines demonstrated his exceptional versatility, achieving immediate success and adding three more 500cc championships to his tally.

Agostini's influence extended beyond his racing achievements to his role as an ambassador for the sport. His charismatic personality and professional approach helped elevate the profile of Grand Prix racing, attracting new fans and sponsors to the championship. His battles with fellow Italian riders like Pier Paolo Bianchi created intense national interest, while his success against international competition demonstrated Italian engineering excellence on a global stage.

The technical knowledge Agostini developed during his career was encyclopedic, covering every aspect of motorcycle performance and setup. His feedback to engineers was precise and actionable, contributing to significant advances in chassis design, suspension setup, and aerodynamic efficiency. This technical expertise made him a valuable test rider even after his competitive career ended, with manufacturers seeking his input on new developments.

Most remarkably, Agostini's career spanned the transition from four-stroke to two-stroke dominance, requiring him to completely adapt his riding techniques and understanding of motorcycle dynamics. His success with both technologies demonstrated an intellectual flexibility and learning capacity that separated him from many of his contemporaries. This adaptability would become a hallmark of truly great champions throughout Grand Prix history.

The legacy of Giacomo Agostini extends far beyond his statistical achievements. He established templates for professionalism, technical development, and competitive excellence that continue to influence modern MotoGP. His approach to fitness, preparation, and mental conditioning became the standard that subsequent generations of riders would emulate. More than five decades after his first championship, Agostini remains the benchmark against which Grand Prix legends are measured.

***

The British motorcycle industry's dominance in the early decades of Grand Prix racing represented more than mere sporting success; it reflected a golden age of engineering innovation and manufacturing excellence that established Britain as the global center of motorcycle development. The achievements of Norton, Triumph, and other British manufacturers during the 1950s and 1960s created a legacy that continues to influence motorcycle design and racing philosophy today.

Mike Hailwood's emergence as the face of British racing excellence perfectly embodied the nation's approach to Grand Prix competition. Born in 1940, Hailwood possessed a natural talent that transcended individual manufacturer loyalties, achieving success aboard Norton, MV Agusta, Honda, and Suzuki machinery throughout his career. His versatility and adaptability made him the perfect representative of an era when rider skill could overcome mechanical disadvantages through sheer determination and technical understanding.

The Norton Manx series represented the pinnacle of British single-cylinder racing development, combining traditional engineering excellence with innovative solutions to the challenges of Grand Prix competition. The Manx's design philosophy emphasized reliability and usability over absolute power, creating machines that allowed riders to maintain consistently high performance throughout race distances. The famous featherbed frame, developed by Rex and Cromie McCandless, provided handling characteristics that were revolutionary for the era and remained competitive long after the engine's power was surpassed by more modern designs.

Norton's approach to racing reflected broader British engineering philosophy, emphasizing practical solutions and proven technology over exotic experimentation. The Manx engine featured a sophisticated single-cylinder design with dual overhead camshafts and advanced valve gear, producing approximately 50 horsepower at 7,000 rpm in 500cc form. While these figures were modest compared to contemporary multi-cylinder machines, the Manx's power delivery was smooth and predictable, allowing riders to maintain higher average speeds through consistent performance.

The development of the featherbed frame represented a breakthrough in motorcycle chassis design that influenced construction techniques for decades. The frame's twin-tube construction provided exceptional rigidity while maintaining relatively low weight, creating a platform that maximized the effectiveness of the engine's power delivery. The geometry was carefully optimized for stability at high speeds while providing responsive handling in technical sections, achieving a balance that many manufacturers struggled to match.

Triumph's racing efforts during this period focused primarily on the larger displacement classes, where their powerful twin-cylinder engines provided competitive advantages. The Triumph racing twins featured advanced engineering solutions including aluminum-bronze cylinder heads, sophisticated carburetion systems, and precision-balanced components that enabled high engine speeds while maintaining reliability. The distinctive sound and character of Triumph racing engines became synonymous with British engineering excellence.

The Triumph approach differed significantly from Norton's single-cylinder philosophy, emphasizing the power advantages of multi-cylinder design while managing the additional complexity through careful engineering and quality control. Triumph racing engines typically produced 10-15% more power than comparable Norton singles, but required more sophisticated maintenance procedures and displayed greater sensitivity to setup parameters.

Mike Hailwood's mastery of these British machines was demonstrated through his nine world championships, achieved across four different displacement classes and with multiple manufacturers. His 1961 season aboard Norton machinery showcased the perfect symbiosis between rider and machine, with victories at the Isle of Man TT and other challenging circuits demonstrating both personal skill and engineering excellence. Hailwood's riding style was characterized by smooth precision and technical understanding that maximized the strengths of British machinery while minimizing inherent weaknesses.

The competitive battles between British and Italian manufacturers during this era created some of the most memorable racing in Grand Prix history. The contrasting philosophies – British practicality versus Italian sophistication – produced machines with distinctly different characteristics that required different riding techniques and tactical approaches. Races often became contests between engineering philosophies as much as rider skill, with outcomes determined by factors ranging from weather conditions to circuit characteristics.

British manufacturers' commitment to racing development extended beyond the premier 500cc class to comprehensive programs across all displacement categories. Norton, Triumph, and other British firms recognized that success in smaller classes provided valuable development opportunities while showcasing engineering capabilities to potential customers. This comprehensive approach created a strong foundation of technical knowledge that benefited all aspects of motorcycle development.

The decline of British dominance in Grand Prix racing during the late 1960s resulted from several converging factors. The emergence of sophisticated Japanese manufacturers with substantial development budgets created competitive pressure that British firms struggled to match. Limited financial resources, combined with changing market conditions and corporate priorities, gradually reduced British manufacturers' ability to compete at the highest level of Grand Prix racing.

However, the technical innovations developed by British manufacturers during their golden years continued to influence motorcycle design long after their competitive decline. Frame construction techniques, engine design principles, and suspension philosophies pioneered by

Norton, Triumph, and other British firms became standard practice throughout the industry. The emphasis on reliability, usability, and balanced performance established during the British era remains fundamental to modern motorcycle development.

The legacy of British racing excellence extends beyond technical achievements to include cultural and sporting contributions that shaped Grand Prix racing's character. The gentlemanly conduct and sportsmanship displayed by riders like Mike Hailwood established behavioral standards that continue to influence the sport's culture. The British approach to racing – combining technical excellence with fair competition and mutual respect – became a template that subsequent generations of riders and manufacturers have emulated.

***

The transformation of Grand Prix motorcycle racing from a primarily European championship to a truly global sport began during the 1960s and accelerated throughout the 1970s, fundamentally changing the competitive landscape while enriching the sport's cultural diversity. This international expansion brought new perspectives, riding techniques, and technical approaches that contributed significantly to the rapid evolution of motorcycle racing during this pivotal period.

Spain's emergence as a motorcycling powerhouse began with Ángel Nieto's dominance in the smaller displacement classes, establishing a Spanish racing tradition that continues to influence MotoGP today. Born in 1947 in Zamora, Nieto's approach to racing was characterized by precision, intelligence, and an almost supernatural ability to extract maximum performance from machinery that was often inferior to his competitors'. His 13 world championships, achieved across the 50cc and 125cc classes between 1969 and 1984, demonstrated that success in Grand Prix racing depended as much on rider skill and dedication as on superior equipment.

Nieto's racing philosophy emphasized consistency and strategic thinking over pure speed, an approach that proved devastatingly effective in championship campaigns. His ability to score points in every race, even when his machinery was not competitive for victories, allowed him to accumulate championship points steadily while rivals suffered from mechanical failures or crashes. This methodical approach to championship competition became a template that subsequent Spanish champions would adopt and refine.

The technical aspects of Nieto's success were equally impressive, as he worked closely with various manufacturers to develop machines that maximized his particular strengths. His feedback to engineers was detailed and practical, focusing on usability and consistency rather than peak performance figures. This collaboration approach contributed to significant advances in small-displacement engine development, particularly in areas such as carburetion, ignition timing, and transmission design.

Spanish manufacturers and sponsors began investing heavily in Grand Prix racing during Nieto's peak years, recognizing the marketing value and technical benefits of international competition. Companies like Bultaco and Derbi developed sophisticated racing programs that challenged established European manufacturers, bringing innovative approaches to engine design and chassis construction. These Spanish firms often emphasized creative engineering solutions over massive development budgets, producing machines that were competitive despite limited resources.

The Australian contribution to Grand Prix racing during this era was personified by riders like Ken Kavanagh and

Tom Phillis, who brought a distinctive antipodean approach to international competition. Australian riders were known for their fearless attacking style and adaptability to different circuits and conditions, qualities that were developed through competition on the challenging Australian racing circuits of the era.

The geographic isolation of Australian racing led to the development of unique technical solutions and riding techniques that proved surprisingly effective in international competition. Australian mechanics and engineers often worked with limited resources, leading to innovative problem-solving approaches and practical modifications that enhanced machine performance. This resourcefulness became a hallmark of Australian involvement in Grand Prix racing, with riders and teams consistently achieving more than their equipment suggested was possible.

American riders began making their presence felt in Grand Prix racing during the late 1970s, bringing a professional approach and aggressive riding style that would eventually revolutionize the sport. While American success would reach its peak in the 1980s, the groundwork was laid during this earlier period by riders who crossed the Atlantic to compete in the world championship despite significant logistical and financial challenges.

The American approach to motorcycle racing was influenced by dirt track and flat track racing traditions that emphasized bike control skills and aggressive riding techniques. These skills proved surprisingly transferable to Grand Prix racing, particularly on circuits that required precise throttle control and the ability to manage sliding or loose-traction situations. American riders often demonstrated superior adaptability to changing track conditions, a skill that became increasingly valuable as the championship expanded to include more diverse circuit types.

Television coverage expansion during this period played a crucial role in globalizing Grand Prix racing, allowing fans around the world to follow their national heroes and develop interest in the championship. The visual appeal of motorcycle racing translated well to television, with dramatic crashes, close racing, and exotic locations providing compelling content for broadcasters. This increased exposure attracted new sponsors and manufacturers to the sport, accelerating its global development.

The technical exchange resulting from international expansion enriched all aspects of Grand Prix development. Manufacturers were exposed to different engineering approaches and solutions, leading to cross-pollination of ideas that accelerated innovation. Riders learned from competitors with different backgrounds and techniques, raising the overall standard of competition while developing more diverse approaches to race craft.

Sponsorship patterns began to change as the championship became more international, with companies from various countries recognizing the global marketing opportunities presented by Grand Prix racing. This increased financial support enabled teams to invest more heavily in development while providing riders with better equipment and support. The professional infrastructure of the sport began to develop rapidly, with specialized transportation, technical support, and media services emerging to serve the growing international community.

The cultural impact of international expansion extended beyond mere competition to include the development of a truly global racing community. Riders, mechanics, and team personnel from different countries worked together, sharing knowledge and techniques while developing friendships and professional relationships that transcended national boundaries. This cosmopolitan atmosphere became one of Grand Prix racing's most appealing characteristics, distinguishing it from more parochial forms of motorsport.

Most importantly, the international expansion of Grand Prix racing during the 1960s and 1970s established the foundation for the global championship that exists today. The inclusion of riders and manufacturers from diverse backgrounds created a competitive environment that demanded constant innovation and improvement, establishing the dynamic that continues to drive MotoGP development. The lessons learned during this pivotal period of expansion continue to influence how the sport approaches new markets and incorporates emerging talent from around the world.

THREE

Technology and Competition (1980-1999)

The 1980s marked the beginning of one of the most intense and productive rivalries in motorsport history, as Yamaha and Honda engaged in a technological arms race that would fundamentally transform Grand Prix motorcycle racing. This competition extended far beyond mere sporting rivalry, representing a clash between two distinct Japanese engineering philosophies that would drive innovation and performance to unprecedented levels.

Kenny Roberts Sr.'s arrival in Grand Prix racing in 1978 as the first American 500cc world champion had already begun to disrupt established European racing traditions, but his continued success aboard Yamaha machinery through 1980 established the foundation for what would become a legendary manufacturer rivalry. Roberts brought an American flat track racing background that emphasized bike control and aggressive riding techniques, skills that proved perfectly suited to the characteristics of Yamaha's two-stroke racing machines.

Yamaha's engineering philosophy during this era emphasized power delivery characteristics and handling balance that complemented aggressive riding styles. The Yamaha YZR500 featured a V4 two-stroke engine that produced exceptional mid-range torque, allowing riders to maintain momentum through technical sections while providing explosive acceleration out of corners. The engine's power characteristics were particularly well-suited to circuits with frequent direction changes, where the ability to accelerate quickly from lower speeds provided significant competitive advantages.

The technical sophistication of Yamaha's approach extended beyond engine design to include comprehensive chassis development and aerodynamic refinement. The YZR500's frame was constructed using advanced aluminum fabrication techniques that provided optimal stiffness-to-weight ratios while allowing precise tuning of handling characteristics for different circuits. Suspension components were developed specifically for the unique demands of two-stroke power delivery, with particular attention to maintaining traction during the violent acceleration and deceleration cycles characteristic of these engines.

Honda's response to Yamaha's early success represented a fundamentally different approach to two-stroke racing technology. The Honda NSR500 featured a unique V3 engine configuration that was technically complex but offered significant advantages in terms of weight distribution and mass centralization. The engine's 120-degree firing intervals provided more consistent power delivery compared to traditional V4 configurations, creating machines that were more predictable and easier to ride at the absolute limit.

Honda's engineering resources and development budget far exceeded those of any competitor, allowing the company to pursue multiple technical solutions simultaneously while conducting extensive testing programs. The NSR500's development involved sophisticated computer modeling and analysis techniques that were revolutionary for the era, enabling Honda engineers to optimize every aspect of machine performance before physical testing began.

The competitive dynamic between Yamaha and Honda created a development spiral that benefited the entire sport.

Each manufacturer's innovations forced the other to respond with even more advanced solutions, leading to rapid progress in all aspects of motorcycle racing technology. Engine power outputs increased dramatically during this period, with the most advanced machines producing over 140 horsepower from 500cc two-stroke engines by the late 1980s.

Freddie Spencer's partnership with Honda during the mid-1980s demonstrated the potential of the NSR500 platform while establishing new standards for rider versatility and professional dedication. Spencer's incredible 1985 season, in which he won championships in both 250cc and 500cc classes, showcased both Honda's technical superiority and the exceptional demands placed on riders competing at the highest level of Grand Prix racing.

Spencer's riding style was characterized by precision and technical perfection that maximized the NSR500's strengths while managing its complexities. His ability to extract consistent performance from the challenging V3 engine configuration required exceptional sensitivity to throttle control and machine setup, skills that he developed through extensive testing and practice sessions. The collaboration between Spencer and Honda's engineers produced significant advances in areas such as chassis geometry, suspension tuning, and aerodynamic development.

The Yamaha-Honda rivalry extended beyond the premier 500cc class to encompass all levels of Grand Prix competition. Both manufacturers developed comprehensive racing programs that included 250cc and 125cc machines, recognizing that success across multiple classes provided valuable technical knowledge while showcasing engineering capabilities to the widest possible audience. The competition in smaller displacement classes often proved even more intense than premier class racing, as closer performance margins made development advantages more decisive.

Technological innovations developed during this rivalry period had profound implications for road motorcycle development. Advances in two-stroke engine design, electronic ignition systems, suspension technology, and aerodynamics were rapidly incorporated into production motorcycles, providing consumers with direct benefits from racing development. The performance and reliability improvements achieved through racing competition significantly enhanced the appeal and capability of road bikes throughout the 1980s.

The psychological aspects of the Yamaha-Honda rivalry were equally important as the technical competition. Riders aligned with each manufacturer developed distinct identities and approaches to racing, with Yamaha riders generally known for aggressive, attacking styles while Honda riders were characterized by technical precision and methodical race craft. These cultural differences added dramatic tension to racing competition while creating compelling narratives for fans and media.

International expansion of the championship during this period provided both manufacturers with global marketing opportunities while exposing their products to diverse racing conditions and technical challenges. Races in different climates and on various circuit types tested machine versatility and reliability, leading to further development advances as manufacturers addressed performance issues revealed by international competition.

The financial investment required to compete effectively in the Yamaha-Honda rivalry established new benchmarks for Grand Prix racing budgets and professionalism. Both manufacturers committed substantial resources to racing development, employing large engineering teams and conducting extensive testing programs that approached

Formula 1 levels of sophistication and expense. This escalation raised competitive standards while potentially excluding smaller manufacturers from serious championship contention.

The Yamaha-Honda rivalry established the template for manufacturer competition that continues to define MotoGP today. The principle that sustained competitive success requires continuous innovation and substantial investment became accepted as fundamental truth, while the benefits of intense rivalry for driving technological progress were clearly demonstrated. The lessons learned during this pivotal period continue to influence how manufacturers approach Grand Prix racing development and competition strategy.

***

The phenomenon of riders competing simultaneously across multiple Grand Prix classes reached its absolute pinnacle with Freddie Spencer's extraordinary 1985 season, when he achieved the seemingly impossible feat of winning world championships in both 250cc and 500cc classes. This accomplishment represented not merely individual brilliance but the culmination of an era when the physical and mental demands of Grand Prix racing, while extreme, had not yet reached the specialized levels that would eventually make such versatility impossible.

Spencer's path to this historic achievement began with his emergence as Honda's premier talent in the early 1980s. Born in 1962 in Louisiana, Spencer possessed a combination of raw speed, technical sophistication, and mental strength that made him ideally suited to Honda's complex and demanding NSR machinery. His initial success in the 500cc class demonstrated exceptional adaptability, as he quickly mastered the unique characteristics of Honda's V3 engine configuration and translated that understanding into race-winning performances.

The technical challenges of competing in multiple classes simultaneously were staggering by any measure. The 250cc and 500cc machines of the mid-1980s were fundamentally different in their power characteristics, handling dynamics, and riding requirements. The Honda NSR250 featured a V2 two-stroke engine that emphasized high engine speeds and precise throttle control, while the NSR500's V3 configuration demanded different techniques for managing its more violent power delivery and greater physical demands.

Spencer's preparation for the 1985 season involved an unprecedented level of physical and mental conditioning that established new standards for professional motorcycle racing. His training regimen included extensive cardiovascular work to build the endurance required for competing in multiple races during Grand Prix weekends, combined with specific exercises designed to develop the muscle memory and reflexes needed to switch between different machines rapidly and effectively.

The logistical complexity of Spencer's dual championship campaign required complete reorganization of Honda's racing department. Separate technical teams were assigned to each class, with Spencer working closely with different engineers and mechanics to optimize setup and performance for each machine. The communication and coordination required to manage this dual effort established precedents for team organization and rider support that influenced Grand Prix racing for decades.

Spencer's 1985 season began with immediate success in both classes, winning the opening 500cc round at Kyalami and following with victory in the 250cc race at the same venue. This early success demonstrated his ability to maintain peak performance levels across different machines while managing the physical and mental stress of competing twice during race weekends. The season's progression required consistent excellence in both classes, as championship points accumulated through sustained performance rather than occasional brilliance.

The mid-season period proved most challenging for Spencer's dual campaign, as the physical demands of frequent racing began to accumulate while mechanical issues with both machines created additional stress. His ability to maintain championship-contending pace in both classes while managing these pressures demonstrated mental resilience that separated him from other talented riders of the era.

The technical development required for Spencer's dual campaign contributed significantly to Honda's understanding of two-stroke racing technology. Data and insights gained from his detailed feedback on both machines provided engineers with comprehensive information about engine performance, chassis dynamics, and rider requirements across different displacement levels. This knowledge accelerated development programs and influenced Honda's approach to racing technology throughout the remainder of the decade.

Spencer's riding technique required constant adaptation as he switched between machines during practice and qualifying sessions. The 250cc bike demanded a more aggressive, high-rev approach that maximized the V2 engine's power band, while the 500cc machine required smoother inputs and more precise throttle control to manage its greater power output. His ability to make these mental and physical adjustments quickly became legendary among fellow competitors and team personnel.

The championship battle intensified during the European rounds, where Spencer faced strong challenges from specialist riders who focused exclusively on single classes. Eddie Lawson's 500cc campaign aboard Yamaha machinery provided fierce competition, while various European riders challenged Spencer's 250cc dominance. The pressure of defending leads in both championships simultaneously created unprecedented stress levels that tested Spencer's mental strength and professional maturity.

The climactic final rounds of the 1985 season saw Spencer secure both championships through methodical, intelligent racing that prioritized points accumulation over spectacular victories. His ability to manage risk while maintaining the pace necessary for championship success demonstrated strategic thinking that was advanced for the era. The achievement established Spencer as one of the greatest riders in Grand Prix history while proving that exceptional talent, properly supported, could overcome seemingly impossible challenges.

The physical toll of Spencer's 1985 campaign was severe, contributing to career-ending injuries that prevented him from defending his championships in subsequent seasons. The demands of competing at the absolute limit across multiple classes exceeded sustainable levels, providing clear evidence that the sport's evolution toward greater specialization was inevitable. Spencer's achievement remained unique partly because the physical and technical demands of Grand Prix racing continued to increase beyond the point where such versatility was possible.

The era of multi-class champions extended beyond Spencer's extraordinary achievement to include other riders who demonstrated exceptional versatility across different displacement categories. Phil Read's success in both 250cc and 500cc classes during the late 1960s and early 1970s established the template for cross-class competition, while riders like Johnny Cecotto and Franco Uncini achieved significant success in multiple categories throughout the 1970s and early 1980s.

The gradual disappearance of multi-class champions reflected the sport's evolution toward greater technical sophistication and physical demands. As machines became more powerful and complex, the specialized knowledge required to extract maximum performance became increasingly detailed and time-consuming to develop. Simultaneously, the physical conditioning required to compete at the highest level became more demanding, making it difficult for riders to maintain peak performance across multiple disciplines.

The legacy of multi-class champions like Spencer continues to influence modern MotoGP through their contributions to rider development and technical understanding. Their comprehensive knowledge of motorcycle dynamics across different displacement levels provided insights that informed subsequent generations of engineers and riders. The professional standards they established for preparation, fitness, and technical collaboration became templates that modern riders continue to follow and refine.

***

The final decades of two-stroke dominance in Grand Prix racing witnessed some of the most spectacular and technically advanced machines ever created, alongside riders who pushed the boundaries of speed and control to their absolute limits. The period from the late 1980s through 2001 represented the culmination of two-stroke development, producing motorcycles that combined devastating power with increasingly sophisticated electronic systems while demanding exceptional skill and bravery from their riders.

Mick Doohan's emergence as the dominant force of the 1990s perfectly embodied the evolution of two-stroke racing toward its ultimate expression. The Australian's partnership with Honda and the NSR500 created one of the most successful combinations in motorsport history, resulting in five consecutive world championships from 1994 to 1998 that established new benchmarks for sustained excellence in Grand Prix racing.

Doohan's approach to racing represented a fundamental shift toward scientific precision and methodical dominance that contrasted sharply with the more instinctive styles of earlier champions. His collaboration with Honda's engineers was extraordinary in its depth and technical sophistication, with every aspect of machine performance subjected to detailed analysis and continuous refinement. This systematic approach to development accelerated the evolution of two-stroke technology while establishing new standards for rider-engineer collaboration.

The Honda NSR500 of Doohan's championship years represented the pinnacle of two-stroke racing development, featuring advances in engine management, chassis design, and aerodynamics that pushed performance to previously unimaginable levels. The engine produced over 180 horsepower in its final iterations, while sophisticated electronic systems provided increasingly precise control over power delivery and handling characteristics. These machines were extraordinarily difficult to ride, requiring exceptional physical strength and mental concentration to manage their violent power characteristics.

Doohan's riding technique was perfectly adapted to the NSR500's demanding characteristics, featuring smooth precision and exceptional consistency that allowed him to extract maximum performance while minimizing the risks associated with such powerful machinery. His ability to maintain race pace while managing tire degradation and fuel consumption demonstrated strategic thinking that was advanced for the era, establishing principles that continue to influence modern racecraft.

The technical challenges of managing 180+ horsepower two-stroke engines were staggering, requiring sophisticated solutions to problems of power delivery, handling stability, and mechanical reliability. Honda's engineers developed increasingly complex electronic systems that could modulate power output based on various parameters, while chassis and suspension components were continuously refined to cope with the enormous forces generated by these engines.

Wayne Rainey's tragic accident at Misano in 1993 marked a watershed moment in Grand Prix racing safety, highlighting the extreme risks associated with the sport's relentless pursuit of speed and performance. Rainey's career had been characterized by exceptional bravery and skill in managing the Yamaha YZR500's challenging characteristics, but his accident demonstrated that even the most talented riders were vulnerable to the consequences of mechanical failure or momentary lapses in concentration.

The safety improvements implemented following Rainey's accident included enhanced medical facilities, improved track safety barriers, and the development of better protective equipment for riders. These changes reflected growing awareness that the sport's continued development required balancing performance advancement with risk management, establishing principles that would become increasingly important as speeds continued to increase.

The competitive landscape of the late two-stroke era featured intense rivalry between Honda, Yamaha, Suzuki, and occasionally other manufacturers, each pursuing different technical solutions to the challenge of extracting maximum performance from 500cc two-stroke engines. Yamaha's V4 configuration emphasized different power characteristics compared to Honda's V3, while Suzuki's square-four engine represented yet another approach to two-stroke racing technology.

These manufacturer rivalries drove rapid development in all areas of motorcycle technology, with innovations in engine management, chassis design, and aerodynamics appearing regularly throughout the racing seasons. The competitive pressure created an environment where technical stagnation meant immediate competitive disadvantage, ensuring continuous advancement in all aspects of machine performance.

The international expansion of Grand Prix racing during this period exposed two-stroke machines to increasingly diverse operating conditions, from the high temperatures of Malaysian rounds to the challenging weather conditions of European circuits. This environmental diversity demanded greater machine versatility while providing valuable development opportunities for manufacturers seeking to improve reliability and performance consistency.

The emergence of electronic traction control and other rider aids during the final years of two-stroke dominance represented early steps toward the sophisticated electronic systems that would characterize four-stroke MotoGP racing. These systems were primitive compared to modern standards but provided valuable lessons about the potential benefits and challenges of electronic performance enhancement.

The rider development systems of this era produced exceptional talent pools that included future legends like Valentino Rossi, who began his Grand Prix career aboard two-stroke machinery in the smaller displacement classes. The progression through 125cc and 250cc two-stroke classes provided comprehensive education in motorcycle dynamics and race craft that prepared riders for the ultimate challenge of 500cc competition.

The economic model of Grand Prix racing during the late two-stroke era established patterns of manufacturer investment and commercial development that continue to influence modern MotoGP. Television coverage expansion and international marketing opportunities attracted increased sponsor investment while raising the profile of the championship among global audiences.

The technical knowledge and experience gained during the final years of two-stroke development provided crucial foundations for the transition to four-stroke MotoGP racing. Engineers and riders who had mastered the complexities of two-stroke technology brought valuable insights to the development of four-stroke machines, ensuring continuity of expertise through the sport's most significant technical transition.

The legacy of the last two-stroke warriors extends beyond their competitive achievements to include their contributions to safety, professionalism, and technical development. Their willingness to compete aboard increasingly powerful and demanding machines pushed the boundaries of human performance while providing valuable lessons about the balance between speed and safety that continues to influence modern racing development.

The end of the two-stroke era in 2001 marked the conclusion of a remarkable period in motorsport history, during which riders and machines achieved performance levels that seemed impossible just decades earlier. The courage, skill, and technical sophistication demonstrated by the last two-stroke warriors established benchmarks for excellence that continue to inspire modern MotoGP competitors while providing lasting contributions to motorcycle racing heritage.

FOUR

The Birth of MotoGP (2000-2006)

The dawn of the new millennium brought with it the most fundamental transformation in Grand Prix motorcycle racing since its inception in 1949. The decision to transition from two-stroke to four-stroke engines wasn't merely a technical adjustment—it was a complete philosophical reimagining of what motorcycle racing could and should be. This monumental shift would not only change the sound, smell, and character of racing but would also create opportunities for manufacturers to develop technologies directly applicable to road motorcycles.

The roots of this transformation can be traced back to the late 1990s when the Fédération Internationale de Motocyclisme began contemplating the future of the premier class. Two-stroke engines, while spectacular in their raw power delivery and distinctive screaming exhaust notes, had become increasingly disconnected from road bike development. Environmental concerns were mounting, and manufacturers were finding it difficult to justify massive investments in technology that had no direct application to their consumer products.

The 500cc two-stroke era had reached extraordinary levels of sophistication by 2001, with machines producing over 200 horsepower from engines weighing less than 65 kilograms. However, these mechanical marvels required such specialized knowledge and manufacturing techniques that only a handful of engineers worldwide truly understood their intricacies. Honda's NSR500, Yamaha's YZR500, and Suzuki's RGV500 represented the absolute pinnacle of two-stroke technology, but they were essentially racing-only curiosities with no relevance to everyday motorcycling.

The FIM's decision to introduce 990cc four-stroke engines for the 2002 season initially allowed both formats to compete simultaneously, creating a fascinating transitional period. The four-stroke machines were given a significant displacement advantage—990cc versus 500cc—to theoretically equalize performance. However, early four-stroke prototypes were significantly heavier and less powerful than their two-stroke counterparts, leading many to question whether the transition would ever truly succeed.

Honda, ever the pioneer, committed fully to the four-stroke concept with their RC211V project. The Japanese manufacturer's engineering team, led by the legendary Shoichiro Irimajiri, approached the challenge with characteristic methodical precision. They recognized that while raw power might initially favor the two-strokes, the superior torque characteristics and fuel efficiency of four-stroke engines would eventually prove decisive.

The RC211V's development represented one of the most intensive engineering projects in Honda's history. The V5 engine configuration was chosen specifically for its compact dimensions and optimal mass centralization, while revolutionary technologies like pneumatic valve actuation—borrowed directly from Formula One—allowed the engine to rev to previously unimaginable speeds for a four-stroke motorcycle engine.

Yamaha initially remained committed to two-stroke development, continuing to refine their YZR500 throughout 2001 and into early 2002. However, as Honda's four-stroke program showed increasing promise, Yamaha quietly began their own four-stroke project. The M1 prototype, first tested in late 2002, utilized a more conventional inline-four configuration but incorporated similarly advanced technologies.

Suzuki and Kawasaki, smaller manufacturers with more limited resources, faced difficult decisions about their premier class futures. Suzuki chose to develop their own four-stroke program, eventually producing the GSV-R, while Kawasaki would later withdraw from the premier class entirely, focusing their efforts on the smaller categories.

The technical regulations governing this transition were carefully crafted to ensure competitive balance while encouraging innovation. The 990cc four-stroke engines were limited to four cylinders, with minimum weight requirements and restrictions on exotic materials. Fuel tank capacity was standardized at 24 liters, creating an additional strategic element as four-stroke engines generally consumed more fuel than their two-stroke predecessors.

The transition marked the beginning of the electronic revolution in motorcycle racing. Four-stroke engines, with their more complex valve trains and combustion characteristics, required sophisticated engine management systems from the outset. These early systems would evolve rapidly, eventually encompassing traction control, wheelie control, and engine braking management—technologies that would fundamentally change riding techniques and competitive dynamics.

The 2002 season became a fascinating laboratory for this technological evolution. Valentino Rossi, riding the final evolution of Honda's NSR500 two-stroke, battled against teammates Alex Barros and Tohru Ukawa on early RC211V four-strokes. The contrast was stark: Rossi's two-stroke was lighter, more agile, and produced devastating acceleration, while the four-strokes offered superior top speed and remarkable consistency over race distances.

By mid-2002, it became clear that the four-stroke evolution was proceeding faster than anyone had anticipated. The RC211V's reliability advantages began to outweigh its performance deficits, and Honda made the strategic decision to switch all their riders to four-stroke machinery for the 2003 season. This decision effectively marked the end of the two-stroke era in premier class racing, as other manufacturers scrambled to accelerate their own four-stroke programs.

The transition's impact extended far beyond the technical specifications. The character of racing changed dramatically as riders adapted to machines with fundamentally different power delivery characteristics. Two-stroke engines provided explosive acceleration but required careful throttle management to maintain traction. Four-strokes offered more linear power delivery but demanded different cornering techniques and braking approaches.

Mechanics and engineers faced steep learning curves as they adapted to the new technology. Carburetor tuning gave way to fuel injection mapping, while traditional two-stroke maintenance procedures became obsolete. The knowledge accumulated over decades of two-stroke development suddenly became historical curiosity rather than competitive advantage.

The sound of racing changed as well. The distinctive, high-pitched scream of two-stroke engines was replaced by the deeper, more guttural roar of four-strokes. While purists lamented this change, the new sound was arguably more accessible to general audiences and certainly more reminiscent of high-performance road motorcycles.

Most importantly, the four-stroke transition established a direct technological connection between racing and road bike development. Manufacturers could now justify their racing investments by pointing to direct applications in their consumer products. Technologies developed for MotoGP began appearing on road bikes within months rather than years, creating a virtuous cycle of innovation that benefited both racing and the broader motorcycling community.

***

No single figure has influenced the modern era of motorcycle racing more profoundly than Valentino Rossi. His transition from promising young talent to global sporting icon coincided perfectly with MotoGP's own transformation, creating a symbiotic relationship that elevated both the sport and its most charismatic champion to unprecedented heights of popular recognition and commercial success.

Rossi's journey to MotoGP supremacy began in the smaller classes, where his natural talent was evident from his first races. Born in Urbino, Italy, in 1979, Rossi inherited his competitive spirit from his father, Graziano Rossi, himself an accomplished motorcycle racer. However, Valentino's approach to racing would prove far more systematic and professional than the previous generation of Italian riders.

His 125cc world championship in 1997 showcased not just raw speed but also tactical awareness beyond his years. Riding for Aprilia, Rossi demonstrated an ability to manage races strategically, understanding when to attack and when to conserve his equipment. This early success was followed by an equally dominant 250cc championship in 1999, again with Aprilia, establishing him as the most promising prospect for premier class success.

The move to 500cc with Honda in 2000 represented a massive step up in complexity and competition. The NSR500 was notoriously difficult to master, with power characteristics that had ended many promising careers. Rossi's first season was a learning experience, finishing second overall behind Kenny Roberts Jr., but his adaptation to the premier class was remarkably quick.

The 2001 season marked Rossi's emergence as the sport's dominant force. His first 500cc world championship was achieved with a combination of raw speed, tactical intelligence, and an increasingly evident flair for the dramatic. Rossi's celebrations after race victories became legendary, from his post-race antics to his elaborate victory celebrations that often incorporated elements of theater and comedy.

But it was Rossi's personality off the motorcycle that truly revolutionized MotoGP's public profile. Previous champions had typically maintained professional but distant relationships with media and fans. Rossi embraced the spotlight with enthusiasm, conducting interviews with wit and charm, engaging with fans directly, and presenting himself as an accessible superstar rather than an aloof athletic machine.

His relationship with the Italian media was particularly significant. Italy had a rich motorcycle racing tradition, but previous Italian champions had often been overshadowed by their international counterparts. Rossi's success coincided with increased television coverage in Italy, creating a perfect storm of national pride and media attention that transformed motorcycle racing into mainstream entertainment.

The transition to four-stroke machinery in 2003 presented Rossi with his greatest technical challenge. The Honda RC211V was fundamentally different from anything he had previously ridden, requiring new techniques and approaches. While some riders struggled with this transition, Rossi adapted with characteristic determination and intelligence.

His 2003 season was a masterclass in adaptation and improvement. Early struggles gave way to increasingly dominant performances as Rossi and his team, led by crew chief Jeremy Burgess, methodically developed both machine and riding technique. The championship victory was never in serious doubt by mid-season, but Rossi continued pushing development, establishing a foundation for future success.

The decision to leave Honda for Yamaha in 2004 shocked the motorcycle racing world. Honda had provided Rossi with championship-winning machinery and seemed poised to continue their dominance. However, Rossi's move was motivated by a desire to prove himself capable of winning with different machinery and to accept Yamaha's technical challenge.

The Yamaha M1 had shown promise but lacked the refinement and outright performance of Honda's RC211V. Rossi's first tests with the machine revealed significant deficiencies in handling and power delivery. However, his feedback and development approach would prove transformative for both rider and manufacturer.

Jeremy Burgess's technical expertise proved crucial during this transition. The Australian engineer had previously worked with Wayne Gardner, Mick Doohan, and other champions, bringing decades of experience to the Yamaha project. Burgess's methodical approach to motorcycle setup and development complemented Rossi's intuitive understanding of vehicle dynamics perfectly.

The 2004 season became one of the most compelling in MotoGP history. Rossi's early victories with Yamaha proved that his Honda success hadn't been solely due to superior machinery. The rivalry with Honda riders, particularly Max Biaggi and later Nicky Hayden, intensified as Rossi systematically dismantled the notion that Honda's technical superiority was insurmountable.

Rossi's riding style evolved during this period, adapting to Yamaha's different characteristics while maintaining his aggressive, overtaking-focused approach. The M1's superior handling in tight corners allowed Rossi to exploit his strengths while minimizing the Honda's straight-line speed advantages.

The psychological warfare that became Rossi's trademark intensified during the Yamaha years. His pre-race comments often contained subtle (and not-so-subtle) attempts to unsettle competitors, while his post-race celebrations became increasingly elaborate. The famous chicken suit worn after his victory at Mugello in 2004 exemplified his ability to combine sporting excellence with entertainment value.

His 2005 championship defense was even more impressive, as competitors had a full year to study and counter Yamaha's approach. Rossi responded by raising his own performance level, producing some of the most memorable rides in MotoGP history. The legendary last-corner pass on Sete Gibernau at Jerez became an instant classic, combining technical skill with perfect timing and psychological dominance.

The development of Rossi's public persona during this period cannot be overstated in its importance to MotoGP's commercial growth. His distinctive yellow color scheme, adopted during his 125cc career, became one of the most recognizable brands in international sport. The "Vale" abbreviation and associated merchandise created revenue streams that dwarfed previous motorcycle racing commercial success.

Rossi's relationship with fans evolved from typical athlete-supporter dynamics to something approaching religious devotion. The "Doctor" nickname, earned through his methodical approach to racing problems, resonated with fans who appreciated both his technical competence and his accessibility. Race weekends became festivals celebrating Rossi's achievements, with yellow-clad supporters creating atmospheric experiences that attracted television viewers worldwide.

His media management was equally revolutionary. Rossi understood that modern sports require entertainment value beyond pure competition. His willingness to engage in verbal sparring with competitors, combined with his natural comedic timing, created storylines that mainstream media could readily embrace. This approach helped elevate MotoGP from niche motorsport to global entertainment property.

The technical evolution of the M1 under Rossi's guidance established Yamaha as Honda's primary competitor, ending years of single-manufacturer dominance. His feedback regarding chassis dynamics, engine characteristics, and electronic systems pushed Yamaha's engineers to innovations that wouldn't have been attempted without his input and championship requirements.

By 2006, Rossi had established himself as not just MotoGP's premier rider but as its primary ambassador and commercial engine. His success had transformed the sport's profile, attracted new fans and sponsors, and created a template for athlete branding that would influence sports marketing far beyond motorcycle racing.

***

The four-stroke revolution of 2002 created unprecedented opportunities for manufacturers to join or rejoin MotoGP's premier class. Unlike the specialized two-stroke era, four-stroke technology offered direct relevance to road bike development, making participation economically justifiable for manufacturers who had previously considered MotoGP too expensive or irrelevant to their core businesses.

Ducati's decision to enter MotoGP represented one of the most significant manufacturer additions in the championship's history. The Italian company had built their reputation on high-performance road bikes featuring distinctive L-twin engines and innovative engineering solutions. However, their racing experience was primarily limited to World Superbike competition, where their 916 and 996 models had achieved considerable success.

The Ducati MotoGP project began in earnest during 2001, with the company's engineers facing the daunting challenge of creating a competitive four-stroke engine while maintaining their distinctive design philosophy. The decision to persist with L-twin engine architecture, rather than adopting the inline-four configuration favored by

Japanese manufacturers, demonstrated remarkable confidence in their engineering approach.

Filippo Preziosi, Ducati's chief engineer for the MotoGP project, approached the challenge with characteristic Italian passion and innovation. The Desmosedici (meaning "sixteen" in Italian, referring to the engine's sixteen valves) project incorporated Ducati's trademark desmodromic valve actuation system, which used mechanical linkages to both open and close valves rather than relying on springs for valve closure.

This system, while more complex and expensive than conventional valve trains, offered significant advantages at the high engine speeds required for MotoGP competition. The precise valve control allowed more aggressive cam profiles and higher rev limits, while the elimination of valve springs reduced reciprocating mass and improved engine response.

Desmosedici's debut in 2003 was met with skepticism from competitors and media alike. The distinctive sound of the L-twin engine was immediately recognizable, but early performance was disappointing. The bike suffered from poor handling characteristics and insufficient power compared to the established Japanese manufacturers.

However, Ducati's development pace was remarkable. By 2004, the Desmosedici had evolved into a genuinely competitive machine, with Loris Capirossi achieving the marque's first MotoGP podium finishes. The bike's unique characteristics—particularly its strong acceleration out of corners—began to suit certain circuits and riding styles.

Suzuki's return to premier class competition after a brief absence represented a more traditional approach to four-stroke development. The Japanese manufacturer had withdrawn from 500cc competition in 2000 but committed to returning with four-stroke machinery for the 2002 season.

The GSV-R project, led by chief engineer Shinichi Sahara, faced the challenge of competing against Honda's significant head start in four-stroke development. Suzuki's approach emphasized their traditional strengths in chassis dynamics and rider-friendly power delivery, rather than pursuing outright power or exotic technologies.

The GSV-R utilized a conventional inline-four engine configuration, similar to Yamaha's approach, but incorporated Suzuki's extensive knowledge gained from their road bike development programs. The engine featured innovative pneumatic valve actuation, borrowed from

Formula One technology, and sophisticated fuel injection systems developed specifically for racing applications.

Kenny Roberts Jr.'s involvement with Suzuki brought significant technical expertise and developmental direction to the project. The American rider's feedback, combined with his father's legendary understanding of motorcycle dynamics, helped accelerate the GSV-R's evolution from promising prototype to race-winning machine.

Suzuki's first MotoGP victory came in 2007 with Chris Vermeulen at Le Mans, validating years of patient development and demonstrating that smaller manufacturers could still compete at the highest level. The victory was particularly significant as it came against established rivals on a track that favored raw power over handling finesse.

Kawasaki's approach to MotoGP participation was notably different from their competitors. The company initially committed to four-stroke development but struggled to match the resources and development pace of larger manufacturers. Their ZX-RR project, while technically competent, never achieved the performance levels necessary for championship contention.

The Kawasaki project was hampered by limited budgets and corporate prioritization of other racing programs. Despite employing talented riders including Alex Hofmann and later Marco Melandri, the team struggled with reliability issues and performance deficits that proved insurmountable.

By 2009, Kawasaki made the difficult decision to withdraw from MotoGP entirely, citing economic pressures and the lack of direct relevance to their road bike programs. This withdrawal highlighted the enormous costs and technical challenges associated with premier class competition, even for established manufacturers.

The entry of new manufacturers fundamentally changed MotoGP's competitive dynamics. The previous Honda-Yamaha duopoly was disrupted by Ducati's innovative approach and Suzuki's traditional engineering excellence. This increased competition forced all manufacturers to accelerate their development programs and explore new technical solutions.

The diversity of engineering approaches became one of MotoGP's most compelling aspects during this period. Honda's V5 engine, Yamaha's inline-four, Ducati's L-twin, and Suzuki's inline-four each represented different philosophical approaches to the same fundamental challenge: creating the fastest motorcycle racing machine possible within the technical regulations.

This technical diversity had profound implications for rider skills and career paths. Riders who excelled on one manufacturer's machine might struggle with a competitor's different characteristics. The ability to adapt to different machines became a crucial skill, separating truly elite riders from those who could only succeed with specific equipment.

The manufacturer competition also drove rapid technological advancement. Each company's engineering team sought competitive advantages through innovation, leading to developments in engine management, chassis construction, aerodynamics, and suspension systems. These innovations often found their way into road bike production, fulfilling the original promise of four-stroke relevance.

The economic impact of increased manufacturer participation was equally significant. Multiple factory teams created more opportunities for riders, mechanics, and engineers, while the increased competition attracted greater media attention and sponsor interest. The sport's commercial value grew substantially as the grid expanded and races became more unpredictable.

The technical regulations evolved to accommodate this increased manufacturer participation while maintaining competitive balance. Rules regarding engine configuration, electronics, and development restrictions were carefully crafted to prevent any single manufacturer from achieving insurmountable advantages while encouraging continued innovation.

By 2006, MotoGP had successfully transformed from a two-manufacturer dominated series to a genuinely multi-manufacturer championship. This transformation not only improved the quality of racing but also established the foundation for the sport's continued growth and commercial success in subsequent decades.

FIVE

MotoGP Conquers the World
(2007-2012)

The period from 2007 to 2012 marked MotoGP's most aggressive expansion phase, as the championship evolved from a primarily European-focused series to a truly global phenomenon. This transformation reflected both the sport's growing commercial appeal and the strategic vision of rights holder Dorna Sports, which recognized that international expansion was essential for long-term growth and sustainability.

The foundation for this global expansion had been laid during the early 2000s, but it was the combination of Valentino Rossi's star power, improved television production values, and the emergence of new markets in Asia and the Americas that created unprecedented opportunities for international growth. The challenge was to maintain the sport's European heritage while embracing new cultures and markets that could provide both commercial value and competitive diversity.

Malaysia's Sepang International Circuit had joined the calendar in 1999, representing an early experiment in Asian expansion. However, the circuit's success in hosting compelling races and attracting significant local interest demonstrated the potential for further Asian market development. The tropical climate created unique challenges for teams and riders, while the enthusiastic local fan base proved that motorcycle racing could transcend cultural boundaries.

The addition of China to the MotoGP calendar in 2005 with the Shanghai International Circuit represented a massive strategic bet on the world's largest potential market. The Chinese Grand Prix faced initial challenges with limited local motorcycle racing culture and regulatory complexities, but it demonstrated Dorna's commitment to establishing MotoGP in emerging markets with enormous long-term potential.

China's inclusion required significant investment in local infrastructure and promotional activities. The absence of a strong motorcycle racing tradition meant that every aspect of the event, from marshal training to media coverage, had to be developed from scratch. However, the enormous potential of the Chinese market justified these investments, particularly as domestic motorcycle manufacturers began showing interest in international competition.

The establishment of the Qatar Grand Prix at the Losail International Circuit in 2004 had already demonstrated the potential for Middle Eastern expansion, but the decision to make this race the season opener and hold it under floodlights created a unique spectacle that captured global attention. The night race concept was revolutionary for motorcycle racing, requiring extensive lighting installations and creating dramatic visual effects that enhanced television coverage worldwide.

Qatar's investment in MotoGP went beyond simply hosting a race. The country's leadership recognized the championship as a vehicle for international prestige and tourism promotion, leading to long-term commitments and facility improvements that established new standards for circuit presentation and spectator amenities. The success of the Qatar night race influenced other circuits to explore similar innovations.

American market development represented perhaps the most significant opportunity and challenge for MotoGP expansion. The United States had a rich motorcycle racing history but had been poorly served by international championship racing since the decline of the traditional American rounds in the 1990s. The return of MotoGP to America required finding suitable circuits and building local interest in a sport that faced competition from established American motorsports.

The decision to race at Mazda Raceway Laguna Seca beginning in 2005 was both inspired and controversial. The California circuit was beloved by American motorcycle racing fans for its spectacular setting and challenging layout, particularly the famous "Corkscrew" corner sequence that provided dramatic overtaking opportunities and spectacular television footage.

However, Laguna Seca's facilities were not originally designed for MotoGP's logistical requirements, necessitating significant upgrades to paddock areas, media facilities, and spectator amenities. The circuit's relatively remote location also limited attendance potential compared to major metropolitan venues.

Despite these challenges, the Laguna Seca races quickly became fan favorites, producing some of the most memorable moments in modern MotoGP history. The 2008 battle between Valentino Rossi and Casey Stoner, featuring Rossi's legendary pass at the Corkscrew, demonstrated the circuit's capacity for generating compelling racing and dramatic moments that resonated with global audiences.

The Indianapolis Motor Speedway's inclusion on the MotoGP calendar from 2008 to 2015 represented an attempt to tap into America's most prestigious motorsports venue and reach mainstream American sports fans. The Indianapolis Grand Prix utilized a modified road course layout that incorporated portions of the famous oval track, creating a unique circuit configuration found nowhere else in motorcycle racing.

Indianapolis brought significant credibility and media attention to MotoGP in the American market. The speedway's association with the Indianapolis 500 and its location in the American Midwest provided access to different demographic groups than the traditional California racing community. The facility's world-class infrastructure also addressed many of the logistical challenges associated with hosting premier international motorsports events.

However, the Indianapolis experiment also highlighted the challenges of adapting MotoGP to American motorsports culture. The circuit layout, while technically challenging, never developed the character or emotional connection that riders and fans experienced at traditional European venues. Surface grip issues and the artificial nature of the road course configuration limited racing quality compared to purpose-built motorcycle racing circuits.

The expansion to Australia with the return of the Phillip Island Grand Prix Circuit to the calendar in 1989 had already proven that MotoGP could succeed in distant markets with strong motorcycle racing traditions. The Australian Grand Prix consistently ranked among the best races of each season, combining spectacular coastal scenery with challenging racing conditions and enthusiastic local support.

Phillip Island's success influenced approaches to other potential markets, demonstrating that geographical isolation need not prevent successful MotoGP events if local enthusiasm and appropriate facilities could be maintained. The circuit's natural setting and demanding layout created racing conditions that tested every aspect of rider and machine performance while providing spectacular visual backdrops for television coverage.

The development of new markets also required significant investment in broadcast infrastructure and promotional activities. Each new territory needed custom media strategies, local broadcasting partners, and promotional campaigns designed to build awareness and understanding of MotoGP among previously unfamiliar audiences.

Language barriers, cultural differences, and varying levels of motorsports infrastructure created unique challenges for each market expansion effort. The success of these initiatives often depended on finding local partners and ambassadors who could bridge the gap between MotoGP's European traditions and local sporting cultures.

The economic impact of global expansion extended beyond direct revenue from race hosting fees and television rights. New markets created opportunities for sponsors seeking international exposure, while manufacturers could justify increased MotoGP investments by pointing to expanded marketing reach in key sales territories.

The expansion period also coincided with improvements in television production technology and global distribution capabilities. High-definition coverage, enhanced graphics packages, and improved commentary teams made MotoGP more accessible to new audiences while maintaining the sport's technical credibility and entertainment value.

However, global expansion also created challenges for traditional European venues and fan bases. Travel costs increased for European fans attending overseas races, while scheduling considerations sometimes forced compromises in traditional race timing and season structure. Balancing global growth with respect for established traditions required careful management and ongoing dialogue with stakeholders.

By 2012, MotoGP had successfully established itself as a genuinely global championship with races on five continents and growing audiences in emerging markets. This expansion created a foundation for sustained commercial growth while exposing new cultures to the unique excitement and technical excellence that defined Grand Prix motorcycle racing.

***

The emergence of Casey Stoner as a championship contender in 2007 represented one of the most remarkable underdog stories in modern motorsport history. The young Australian's partnership with Ducati not only ended

Valentino Rossi's seemingly unbreakable dominance but also validated the Italian manufacturer's unconventional approach to MotoGP competition, proving that innovation and determination could overcome resource disadvantages and established hierarchies.

Stoner's path to MotoGP success was far from conventional. Born in Southport, Australia, in 1985, he began his racing career in dirt track competition before transitioning to road racing as a teenager. His family's financial limitations meant that European racing opportunities came only through significant personal sacrifice and the support of dedicated sponsors who recognized his exceptional natural talent.

The move to European competition in 2000 as a 15-year-old represented a massive gamble for the Stoner family. Unlike riders from established racing nations, Stoner lacked access to well-funded development programs or clear pathways to premier class competition. His success in 125cc and 250cc categories came through raw speed and determination rather than sophisticated technical support.

Stoner's riding style was evident from his earliest European races. His ability to maintain high corner speeds while managing sliding motorcycles seemed almost supernatural, suggesting an innate understanding of motorcycle dynamics that couldn't be taught or coached. This natural ability would prove perfectly suited to Ducati's unique characteristics.

The transition to MotoGP in 2006 with LCR Honda provided Stoner with his first taste of premier class competition, but the satellite team's limited resources prevented him from demonstrating his full potential. However, his occasional flashes of brilliance caught the attention of Ducati's management, who were seeking a rider capable of maximizing their innovative but challenging Desmosedici's potential.

Ducati's approach to MotoGP had always been unconventional. While Japanese manufacturers pursued incremental improvements to proven concepts, Ducati's engineers embraced radical solutions that reflected their road bike philosophy. The Desmosedici's L-twin engine configuration, desmodromic valve actuation, and carbon fiber chassis represented a completely different approach to premier class competition.

The Desmosedici's characteristics were both its greatest strength and most significant weakness. The engine's massive torque production and distinctive power delivery provided incredible acceleration out of corners, while the chassis's rigidity offered precise handling feedback. However, these same characteristics made the bike exceptionally difficult to ride, with few riders capable of exploiting its potential while managing its challenging behavior.

Previous Ducati riders had struggled to adapt to Desmosedici's demands. The bike's tendency to slide both front and rear wheels simultaneously required constant corrections and an aggressive riding style that exhausted most riders over race distances. Additionally, the engine's abrupt power delivery made precise throttle control extremely difficult, particularly in changing track conditions.

Stoner's signing with Ducati for the 2007 season was met with skepticism from most observers. The Australian was still largely unknown outside hardcore racing circles, while Ducati's previous results suggested that the Desmosedici was fundamentally flawed as a championship contender. The partnership seemed destined to disappoint both parties' ambitions.

However, the combination of Stoner's unique riding ability and Desmosedici's distinctive characteristics created an almost magical synergy. Stoner's natural ability to manage sliding motorcycles at high speed perfectly complemented the Ducati's tendency to move around beneath its rider. Rather than fighting the bike's behavior, Stoner learned to use its characteristics as competitive advantages.

The 2007 season opened with Stoner's victory at Qatar, immediately serving notice that the Australian-Ducati combination would be a factor in championship competition. However, few observers recognized this as the beginning of a dominant campaign rather than an isolated success on a circuit that might have suited Desmosedici's characteristics.

As the season progressed, Stoner's consistency and speed became impossible to ignore. His victory at the Spanish Grand Prix at Jerez, traditionally one of Valentino Rossi's strongest circuits, demonstrated that the Ducati's competitiveness wasn't limited to specific track types or conditions. The Australian's methodical accumulation of points suggested a championship challenge that few had anticipated.

Stoner's riding style during this period was revolutionary in its aggression and precision. The Desmosedici required constant input from its rider, with traditional smooth riding techniques proving ineffective. Stoner developed a technique that combined aggressive body positioning with precise throttle control, allowing him to maintain high speeds while managing the bike's challenging behavior.

The technical partnership between Stoner and Ducati's engineers, led by Filippo Preziosi, proved crucial to their success. Stoner's feedback helped refine Desmosedici's setup philosophy, while the engineers provided solutions to specific handling issues that had previously limited the bike's potential. This collaboration represented a perfect match between rider intuition and engineering innovation.

The psychological impact of Stoner's early success extended far beyond championship standings. Valentino Rossi, accustomed to dominating MotoGP competition, found himself genuinely challenged for the first time since his early Honda days. The Italian's attempts to match Stoner's pace often resulted in crashes or poor results, suggesting that the Ducati-Stoner combination had genuinely shifted competitive dynamics.

Stoner's championship victory in 2007 was achieved with a combination of raw speed, tactical intelligence, and remarkable consistency. His ten race victories from eighteen rounds demonstrated both Desmosedici's potential and his own ability to extract maximum performance across diverse circuit types and conditions. The championship was effectively secured with several rounds remaining, highlighting the dominance of this unexpected partnership.

The impact of Stoner's success extended beyond individual achievement to validate Ducati's entire MotoGP philosophy. The Italian manufacturer had proven that unconventional engineering approaches could succeed at the highest level of motorcycle racing, inspiring continued innovation and justifying their substantial investment in premier class competition.

Stoner's riding technique influenced an entire generation of young riders who studied his methods for managing difficult motorcycles. His ability to maintain speed while sliding both wheels became a template for riding modern MotoGP machines, which increasingly required aggressive techniques to extract maximum performance.

The commercial impact of the 2007 championship was equally significant for Ducati. The success generated enormous media attention and marketing value, while sales of road bikes benefited from the association with championship success. The Desmosedici's victories provided credibility for Ducati's entire product range, justifying the racing investment through increased brand value and market position.

However, Stoner's championship also highlighted the challenges of sustaining success in MotoGP's rapidly evolving environment. Other manufacturers quickly began studying Ducati's approach and developing counter-strategies, while technical regulations evolved in ways that sometimes disadvantaged the Desmosedici's unique characteristics.

The 2008 season proved that Stoner and Ducati's success wasn't a fluke, as the combination remained competitive throughout the campaign. However, Rossi's resurgence with Yamaha and Honda's continued development demonstrated that sustained championship success required constant evolution and adaptation.

By 2009, the competitive balance had shifted once again, with Ducati struggling to maintain the performance advantages that had enabled their 2007 triumph. This evolution reflected MotoGP's fundamental characteristic: no advantage, however significant, remains permanent in the face of determined competition and unlimited innovation.

***

The emergence of Jorge Lorenzo as a championship contender represented the arrival of a new generation of riders who approached MotoGP with scientific precision and psychological sophistication that surpassed even their accomplished predecessors. Lorenzo's methodical approach to racing, combined with his distinctive personality and unwavering self-confidence, created one of the most compelling and controversial figures in modern motorcycle racing history.

Born in Palma, Mallorca, in 1987, Lorenzo's early career was shaped by the Spanish motorcycle racing boom that had begun in the 1990s with riders like Àlex Crivillé and Carlos Checa achieving international success. However, Lorenzo's approach to racing development was notably different from his compatriots, emphasizing technical precision and mental preparation over pure aggression or natural feel.

Lorenzo's progression through the junior categories was marked by both exceptional speed and frequent crashes as he learned to balance his ambitious overtaking attempts with the need for consistent point scoring. His 250cc world championship in 2006 and 2007 demonstrated remarkable improvement in race management and tactical awareness, suggesting that his transition to MotoGP would be more successful than many young riders who relied primarily on raw speed.

The move to Yamaha's factory team in 2008 created immediate tension within the organization, as Lorenzo was positioned as Valentino Rossi's heir apparent despite having no premier class experience. This situation was unprecedented in modern MotoGP, where established champions typically enjoyed unchallenged status within their teams and expected newcomers to serve supporting roles during their learning phases.

Lorenzo's approach to this challenging situation demonstrated the mental strength that would define his entire career. Rather than being intimidated by Rossi's reputation and achievements, the young Spaniard embraced the opportunity to learn from the master while maintaining his own ambitious goals. This confidence, which some observers interpreted as arrogance, proved essential to Lorenzo's rapid development and eventual success.

The technical challenges of adapting to MotoGP machinery were compounded by the need to work alongside Rossi's established technical crew and development programs. The Yamaha M1 had been specifically developed to suit Rossi's riding style and preferences, creating potential conflicts with Lorenzo's different approach to motorcycle setup and handling characteristics.

Lorenzo's riding style was notably different from Rossi's intuitive and adaptive approach. The Spaniard preferred precisely calibrated machine settings and consistent track conditions, allowing him to implement his methodical approach to corner entry and exit techniques. This precision-oriented style would prove devastatingly effective but required more specific setup parameters than Rossi's more flexible approach.

The 2008 season provided Lorenzo with essential learning experiences while demonstrating his remarkable potential. His first MotoGP victory at Estoril came through a combination of perfect race strategy and flawless execution, showcasing the tactical awareness that had characterized his 250cc championships. However, several crashes while attempting to match Rossi's pace highlighted the learning curve required for premier class success.

Lorenzo's development during his rookie season was accelerated by access to Yamaha's exceptional technical resources and the competitive pressure created by Rossi's presence. The internal rivalry pushed both riders to higher performance levels while providing Lorenzo with immediate feedback on his progress relative to the sport's established benchmark.

The psychological dynamics within the Yamaha team became increasingly complex as Lorenzo's speed and confidence grew. His willingness to challenge Rossi's setup directions and technical preferences created productive tension that ultimately benefited both riders' development. However, this assertiveness also generated media speculation about team harmony and internal politics.

Lorenzo's breakthrough 2009 season demonstrated that his 2008 performances weren't beginner's luck but evidence of genuine championship potential. His four victories and consistent podium finishes established him as Rossi's primary challenger while proving that Yamaha's M1 could be adapted to suit different riding styles without compromising its fundamental strengths.

The Spanish media attention surrounding Lorenzo's success created additional pressures and opportunities that he managed with remarkable maturity. His articulate interviews and willingness to discuss technical aspects of racing helped build his public profile while establishing him as an intellectual approach to competition that appealed to sophisticated audiences.

Lorenzo's 2010 championship campaign represented the culmination of three years of methodical development and preparation. The season-long battle with Rossi, Dani Pedrosa, and Casey Stoner produced some of the most compelling racing in MotoGP history while demonstrating Lorenzo's ability to perform under maximum pressure.

The turning point of the 2010 season came with Rossi's leg injury at Mugello, which effectively ended the Italian's championship challenge while providing Lorenzo with a psychological advantage that he maintained throughout the remainder of the campaign. However, Lorenzo's title wasn't simply the result of his rival's misfortune—his consistency and speed throughout the season demonstrated genuine championship quality.

Lorenzo's riding technique during this period was characterized by exceptional smoothness and precision. His ability to maintain identical lap times throughout race distances while preserving tire performance set new standards for consistency in MotoGP. This approach proved particularly effective during an era when tire management became increasingly critical to race success.

The mental aspects of Lorenzo's preparation were equally impressive. His pre-race routines, developed with sports psychologist support, included detailed visualization exercises and specific rituals designed to optimize his mental state for competition. This scientific approach to psychological preparation influenced other riders and established new standards for professional mental training.

Lorenzo's celebration style became as distinctive as his riding technique. His theatrical post-victory performances, including elaborate costumes and choreographed routines, reflected both his Spanish cultural background and his understanding of modern sports entertainment requirements. These celebrations generated significant media attention while building his personal brand.

The technical evolution of the M1 under Lorenzo's influence was significant. His feedback regarding chassis stiffness, suspension characteristics, and electronic system calibration pushed Yamaha's engineers toward setup philosophies that would influence their bike development for years. This technical contribution demonstrated that Lorenzo's value extended beyond pure riding ability to include genuine developmental insight.

Lorenzo's relationship with media and fans evolved considerably during his championship years. Initial perceptions of arrogance gave way to appreciation for his honesty, technical knowledge, and commitment to excellence. His willingness to admit mistakes and discuss his psychological challenges humanized him while maintaining his competitive credibility.

The impact of Lorenzo's success on Spanish motorcycle racing was profound. His championship validated Spain's investment in rider development programs while inspiring a new generation of young Spanish riders. The commercial success of motorcycle racing in Spain increased significantly during Lorenzo's peak years, creating sustainable foundations for continued Spanish involvement in international competition.

Lorenzo's training methods became legendary throughout the paddock. His physical conditioning program, developed with professional trainers, emphasized functional strength and cardiovascular endurance specific to motorcycle racing demands. This scientific approach to physical preparation raised standards throughout the grid and influenced training methodologies across all categories.

The psychological warfare that developed between Lorenzo and his competitors was more subtle than previous generations' approaches but equally effective. His consistent public confidence, combined with methodical preparation and flawless execution, created doubt in rivals' minds about their own capabilities and preparation levels.

By 2012, Lorenzo had established himself as the premier example of scientific approach to motorcycle racing success. His combination of raw talent, methodical preparation, and unwavering mental strength created a template for modern professional racing that influenced riders, teams, and manufacturers throughout the sport.

SIX

Dominance Redefined (2013-2019)

The arrival of Marc Marquez in MotoGP represented more than just another talented rookie joining the premier class—it marked the beginning of a paradigm shift that would fundamentally alter how motorcycle racing success was measured and achieved. Marquez's immediate dominance upon entering MotoGP in 2013 shattered conventional wisdom about learning curves and established new benchmarks for what was possible in premier class competition.

Born in Cervera, Spain, in 1993, Marquez emerged from the Spanish motorcycle racing system that had been revitalized by earlier champions like Àlex Crivillé and Jorge Lorenzo. However, his approach to racing development was distinctly different from his predecessors, characterized by an almost reckless willingness to explore the absolute limits of adhesion and control that occasionally resulted in spectacular crashes but more often produced impossible-seeming saves and unprecedented lap times.

Marquez's junior career progression was meteoric by any standard. His 125cc world championship in 2010 at age 17 demonstrated exceptional racecraft and tactical awareness, while his dominance in the Moto2 category during 2012 suggested that his talents would translate effectively to four-stroke machinery. However, no previous junior category success had adequately predicted the revolutionary impact he would have upon reaching MotoGP.

The decision to promote Marquez directly to the Honda factory team for 2013 was controversial within the paddock. Traditional wisdom suggested that rookie riders required development time with satellite teams before being entrusted with factory equipment and championship expectations. Honda's willingness to place their faith in an unproven MotoGP rider reflected both their confidence in his abilities and the strategic necessity of finding a replacement for the aging Dani Pedrosa as their primary championship contender.

Marquez's first MotoGP test sessions immediately suggested that conventional expectations might prove inadequate. His lap times were competitive from his earliest runs, while his riding style displayed the aggressive commitment to limit-finding that would define his entire career. The combination of raw speed and apparent fearlessness created immediate speculation about his championship potential.

The RC213V that Marquez inherited for his rookie season represented Honda's most advanced racing motorcycle, incorporating years of development experience and technological innovation. However, the bike's characteristics had been specifically tailored to suit Dani Pedrosa's smooth, precise riding style, creating potential conflicts with Marquez's more aggressive approach to motorcycle control.

Rather than adapting his riding style to suit the motorcycle, Marquez worked with Honda's engineers to modify the RC213V's setup and characteristics to accommodate his unique techniques. His feedback regarding chassis stiffness, suspension calibration, and electronic system parameters pushed Honda's development in new directions that ultimately benefited the bike's overall performance envelope.

Marquez's second race victory at the Circuit of the Americas demonstrated that his early success wasn't circuit-specific or weather-dependent. The dominant performance, achieved against established champions including Lorenzo, Rossi, and Pedrosa, served notice that the competitive hierarchy was undergoing fundamental disruption.

The technical aspects of Marquez's riding technique were revolutionary in their apparent contradiction of basic motorcycle dynamics principles. His ability to save seemingly impossible front-end slides while maintaining competitive lap times defied conventional understanding of tire adhesion limits and motorcycle control techniques. High-speed photography and telemetry analysis revealed methods that other riders struggled to understand, let alone replicate.

Marquez's approach to corner entry was particularly distinctive. While conventional technique emphasized smooth transitions and progressive lean angle increases, Marquez often entered corners with abrupt direction changes and maximum lean angles that should have resulted in immediate crashes. His ability to maintain control under these conditions suggested either exceptional natural ability or revolutionary understanding of motorcycle dynamics.

The psychological impact of Marquez's early success extended throughout the MotoGP paddock. Established champions who had dominated the sport for years suddenly found themselves struggling to match the pace of a rookie rider who seemed immune to the pressure and learning curve that typically defined premier class adaptation periods.

Valentino Rossi, entering the twilight of his legendary career, was particularly affected by Marquez's emergence. The Italian master's attempts to match the young Spaniard's aggressive techniques often resulted in crashes or poor results, highlighting the generational transition occurring within the sport. The contrast between Rossi's calculated approach and Marquez's apparent fearlessness symbolized the changing nature of MotoGP competition.

Jorge Lorenzo's response to Marquez's challenge was equally revealing. The precise Spaniard's methodical approach to racing had served him well throughout his career, but Marquez's ability to find speed through seemingly impossible techniques forced Lorenzo to reconsider fundamental aspects of his riding philosophy. The internal Yamaha dynamics became increasingly complex as Lorenzo struggled to match Honda's new weapon.

Marquez's first championship victory in 2013 was achieved with a statistical dominance that exceeded even the most optimistic predictions. His six victories from eighteen rounds might not have represented the highest win percentage in championship history, but the manner of those victories—often achieved through late-race charges or impossible saves—created a mythology around his abilities that transcended mere numbers.

The rookie championship was unprecedented in the four-stroke era and hadn't been achieved since Kenny Roberts Sr. 's 1978 triumph in the two-stroke 500cc category. However, the modern MotoGP grid's depth and technical sophistication made Marquez's achievement even more remarkable than his historical precedent.

Marquez's celebration style reflected both his youth and his Spanish cultural background. His exuberant post-victory demonstrations and willingness to engage with fans and media created an immediate connection with audiences worldwide. Unlike some previous champions who maintained professional distance, Marquez embraced the entertainment aspects of modern sports while never compromising his competitive focus.

The commercial impact of Marquez's success was immediate and substantial. Spanish television audiences reached unprecedented levels for motorcycle racing, while merchandise sales and sponsor interest increased dramatically. His appeal to younger demographics helped expand MotoGP's audience base beyond traditional motorcycle racing enthusiasts.

Honda's investment in Marquez quickly proved justified through both championship success and technological advancement. His aggressive riding style and willingness to push beyond conventional limits provided valuable feedback for motorcycle development programs, while his commercial appeal enhanced Honda's marketing value across global markets.

The 2014 season demonstrated that Marquez's rookie success wasn't anomalous, as he defended his championship with even greater dominance. His thirteen victories from eighteen rounds established new benchmarks for single-season success while proving that his methods could be sustained over multiple campaigns.

Marquez's approach to race strategy was as revolutionary as his riding technique. Rather than managing races conservatively to ensure point-scoring positions, he consistently pursued maximum attack throughout entire race distances. This approach occasionally resulted in crashes, but more often produced victories that demoralized competitors and accumulated championship points at unprecedented rates.

The technical development of the RC213V accelerated under Marquez's influence. His feedback regarding engine characteristics, chassis behavior, and electronic system requirements pushed Honda's engineers toward innovations that wouldn't have been considered without his unique input. The bike's evolution during this period reflected the symbiotic relationship between exceptional rider ability and advanced engineering support.

Marquez's influence on riding technique extended throughout the MotoGP grid and into lower categories. Young riders studied his methods through video analysis and attempted to replicate his aggressive approach to limit-finding, though few possessed the natural ability to implement his techniques successfully. The "Marquez style" became a recognized approach to motorcycle racing that influenced an entire generation.

The safety implications of Marquez's riding style generated considerable discussion within the paddock and among safety officials. His frequent saves from seemingly impossible situations were spectacular to watch but raised questions about the long-term sustainability of such aggressive techniques. However, his crash rate was actually lower than many competitors, suggesting that his methods, while visually dramatic, were actually more controlled than they appeared.

By 2016, Marquez had established himself not just as MotoGP's dominant force but as a transformative figure who had fundamentally altered the sport's competitive landscape. His success had forced competitors, manufacturers, and even rule-makers to reconsider basic assumptions about motorcycle racing, creating ripple effects that would influence the sport for years to come.

***

The period from 2013 to 2019 witnessed the most intensive technological development phase in MotoGP history, as manufacturers pursued every possible advantage through increasingly sophisticated electronic systems and revolutionary aerodynamic innovations. This arms race was driven by both the competitive pressure created by Marc Marquez's dominance and the continued evolution of four-stroke engine technology, which had reached performance levels that demanded equally advanced control systems.

The electronic revolution in MotoGP had begun during the early four-stroke era, but the 2010s saw these systems evolve from simple engine management tools to comprehensive vehicle control networks that managed every aspect of motorcycle performance. Traction control, which had been relatively primitive during the 2000s, became extraordinarily sophisticated, using wheel speed sensors, lean angle measurements, and throttle position data to prevent rear wheel spin while maximizing acceleration.

Honda's approach to electronics development was characteristically methodical and comprehensive. Their engineers, working closely with Marc Marquez's feedback, created systems that could adapt to his aggressive riding style while providing the safety margins necessary for sustained championship competition. The RC213V's electronics package became the benchmark against which all other systems were measured.

The traction control systems of this era were far more complex than their earlier incarnations. Rather than simply cutting power when wheel spin was detected, these systems could modulate engine output with incredible precision, allowing riders to maintain maximum acceleration while keeping rear wheel slip within optimal parameters. The systems could differentiate between beneficial slip that enhanced forward drive and detrimental spin that wasted power and caused tire degradation.

Yamaha's electronics philosophy differed significantly from Honda's approach. Their engineers emphasized rider-friendly systems that provided assistance without overwhelming the rider's natural feel for the motorcycle's behavior. Jorge Lorenzo's feedback was instrumental in developing this philosophy, as his riding style demanded precise control over electronic intervention levels.

The introduction of wheelie control systems represented another significant technological advancement. These systems used sophisticated algorithms to detect when the front wheel was lifting and would modulate engine power or adjust engine braking to maintain optimal weight distribution. The technology was particularly crucial for maximizing acceleration out of corners, where wheelie control could allow full throttle application much earlier than would otherwise be possible.

Engine braking management became equally sophisticated during this period. Modern four-stroke engines produced significant compression braking that could unsettle motorcycle handling during corner entry. Electronic systems were developed to manage this engine braking through precise fuel injection control and throttle butterfly positioning, allowing riders to maintain stability while maximizing braking performance.

The seamless transmission technology introduced by Honda represented perhaps the most significant mechanical innovation of this era. Traditional motorcycle transmissions created power interruptions during gear changes that affected both acceleration and handling. Honda's seamless system eliminated these interruptions by using complex mechanical and electronic systems to maintain power delivery throughout the shifting process.

This transmission technology provided measurable advantages in lap times and race performance, as riders could maintain maximum acceleration without the traditional compromises associated with gear changes. The system's complexity required extensive software development and mechanical precision that demonstrated Honda's commitment to technological leadership.

Ducati's response to Honda's seamless transmission was characteristically innovative. Rather than copying Honda's mechanical approach, Ducati developed electronic solutions that minimized the impact of traditional transmission design limitations. Their quick-shifter technology became remarkably sophisticated, allowing gear changes to occur with minimal power interruption even with conventional transmission architecture.

The aerodynamic revolution that began during this period was equally dramatic. While motorcycle aerodynamics had traditionally focused on drag reduction and top speed optimization, manufacturers began exploring downforce generation to improve cornering and braking performance. This development was inspired by Formula One technology but required adaptation to the unique challenges of motorcycle racing.

Ducati pioneered the use of aerodynamic winglets on their Desmosedici during the 2015 season. These small appendages, mounted on the fairing sides, generated downforce that pressed the front wheel more firmly onto the track surface. This additional downforce improved braking stability and allowed more aggressive cornering angles, providing measurable performance advantages.

The visual impact of aerodynamic devices was immediate and controversial. Traditional motorcycle racing aesthetics emphasized clean, streamlined designs, while the new winglets created industrial appearances that some observers found unappealing. However, the performance benefits were undeniable, forcing other manufacturers to develop their own aerodynamic solutions.

Honda's aerodynamic development took a different approach, integrating downforce-generating elements into the overall fairing design rather than using external winglets. This philosophy maintained a more traditional visual appearance while achieving similar aerodynamic benefits. The integration required extensive wind tunnel testing and computational fluid dynamics analysis to optimize both performance and appearance.

Yamaha initially resisted the aerodynamic trend, believing that their chassis advantages could compensate for any aerodynamic deficits. However, as other manufacturers' advantages became apparent, Yamaha was forced to develop their own aerodynamic packages. Their approach emphasized subtle design modifications that generated downforce without dramatically altering the M1's traditional appearance.

The regulatory response to aerodynamic development was complex and evolving. The FIM initially allowed relatively free development, recognizing that aerodynamic innovation was driving technological advancement that could benefit the broader motorcycle industry. However, safety concerns about winglets breaking during crashes led to stricter regulations regarding attachment methods and materials.

The 2017 introduction of standardized electronics marked a fundamental shift in MotoGP's technological development. The decision to mandate Magneti Marelli's unified software package eliminated manufacturer advantages in electronics while reducing development costs and increasing competitive parity. This change was controversial but necessary to prevent electronics development from overwhelming other aspects of motorcycle racing.

The transition to spec electronics created new challenges for manufacturers and riders. Teams that had relied on superior electronics to mask other deficiencies suddenly found their competitive positions compromised, while manufacturers with strong chassis and engine packages could exploit these advantages more effectively. The adjustment period was lengthy and affected different teams in various ways.

Riders were forced to adapt their techniques to the standardized electronics, which behaved differently from the manufacturer-specific systems they had learned to exploit. Marc Marquez's adaptation was particularly impressive, as he maintained his dominance despite losing the sophisticated Honda electronics that had partially enabled his aggressive riding style.

The impact of these technological developments extended beyond pure performance to influence riding techniques and race strategies. Riders could attack more aggressively knowing that electronic systems would prevent catastrophic mistakes, while race strategies evolved to account for the consistent performance enabled by advanced electronics.

The commercial implications of this technological arms race were significant. Manufacturers justified enormous development expenses by pointing to road bike applications for racing-derived technologies. Traction control, ABS, and other electronic systems developed for MotoGP began appearing on production motorcycles within months of their racing debuts.

However, the escalating costs of technological development became a concern for smaller manufacturers and teams. The expense of remaining competitive in the electronics and aerodynamics races threatened to exclude participants who couldn't match the resources of larger manufacturers, potentially reducing grid diversity and competitive balance.

The introduction of cost control measures became increasingly necessary to maintain sustainable competition. Regulations limiting testing, standardizing certain components, and restricting development timelines were implemented to prevent technological advancement from completely overwhelming competitive balance and financial sustainability.

By 2019, MotoGP had achieved unprecedented levels of technological sophistication while maintaining competitive excitement and safety standards. The balance between innovation and regulation had created a framework that encouraged advancement while preventing any single technological area from dominating competition completely.

***

While Marc Marquez's dominance defined the era from 2013 to 2019, the period's most compelling narratives often centered on the riders who challenged his supremacy and created the competitive tension that made this era memorable. Andrea Dovizioso's late-career renaissance with Ducati and Maverick Viñales' emergence as Yamaha's standard-bearer represented the highest levels of premier class competition, producing championship battles and individual race performances that rank among MotoGP's greatest moments.

Andrea Dovizioso's transformation from capable midfielder to championship contender represents one of the most remarkable career evolutions in modern motorcycle racing. The Italian rider's early MotoGP career with Honda had been characterized by consistent but unremarkable performances, suggesting that he lacked the exceptional speed necessary for championship success. However, his move to Ducati in 2013 coincided with both his personal maturation as a rider and the Italian manufacturer's technological breakthrough that would make them genuinely competitive with Honda and Yamaha.

Dovizioso's riding style was perfectly suited to the Ducati Desmosedici's unique characteristics. While other riders struggled with the bike's aggressive power delivery and challenging handling, Dovizioso's smooth, calculated approach allowed him to maximize the machine's strengths while minimizing its weaknesses. His ability to manage tire wear over race distances became legendary, often allowing him to mount late-race charges when competitors were struggling with degraded tire performance.

The development of Dovizioso's late-braking technique became one of the most distinctive aspects of his riding during this period. His willingness to brake later and deeper into corners than his competitors often created overtaking opportunities that seemed impossible to other observers. This technique required not only exceptional skill but also supreme confidence in the Ducati's braking capabilities and his own judgment of adhesion limits.

The 2017 season marked Dovizioso's emergence as a genuine championship contender. His six victories throughout the campaign demonstrated both his speed and consistency, while his season-long battle with Marc Marquez produced some of the most memorable races in recent MotoGP history. The championship wasn't decided until the final round, with Dovizioso maintaining mathematical possibilities until the season's closing stages.

The psychological warfare between Dovizioso and Marquez during their championship battles was subtle but intense. Dovizioso's calm, methodical approach contrasted sharply with Marquez's aggressive style, creating compelling narrative tension that extended throughout multiple seasons. Their on-track battles were characterized by mutual respect and exceptional racecraft, producing overtaking sequences that became instant classics.

Dovizioso's relationship with Ducati's engineering team, led by Gigi Dall'Igna, was crucial to both the rider's success and the manufacturer's technological development. His feedback regarding chassis behavior, engine characteristics, and aerodynamic effects helped guide Ducati's development programs toward solutions that not only suited his riding style but also created a more versatile platform for other riders.

The Austrian Grand Prix battles between Dovizioso and Marquez in 2017, 2018, and 2019 became legendary examples of premier class racing at its finest. The Red Bull Ring's characteristics seemed to equalize the competitive differences between Ducati and Honda machinery, allowing pure riding skill to determine race outcomes. These battles showcased both riders' exceptional abilities while demonstrating the narrow margins that separate victory from defeat at MotoGP's highest levels.

Maverick Viñales' arrival at Yamaha in 2017 created enormous expectations for the Spanish rider who had shown flashes of brilliance during his Suzuki years. His early performances with the M1 suggested that he possessed the speed necessary to challenge Marquez regularly, with several dominant victories demonstrating his potential to become Yamaha's first consistent championship contender since Jorge Lorenzo's departure.

Viñales' riding style was characterized by exceptional corner speed and smooth technique that seemed ideally suited to the Yamaha's chassis characteristics. His ability to carry high speeds through long corner sequences often created advantages that competitors couldn't match, particularly on circuits with flowing layouts that rewarded precision over raw power.

However, Viñales' career was also marked by frustrating inconsistency that prevented him from sustaining championship challenges. His tendency to struggle with tire management during races often negated qualifying performances that demonstrated his outright pace. The psychological pressure of championship expectations seemed to affect his race performance, creating cycles of promise and disappointment that defined his Yamaha tenure.

The technical relationship between Viñales and Yamaha's engineers proved more complex than initially anticipated. His setup preferences and feedback sometimes conflicted with the development direction established by previous Yamaha champions, creating challenges for engineers attempting to optimize the M1's performance characteristics. These technical difficulties often manifested as performance inconsistencies that limited his championship potential.

Viñales' occasional dominant performances demonstrated his genuine speed and potential. His victories often came through controlling races from the front, using his exceptional qualifying pace to establish early leads that he could maintain throughout race distances. These performances suggested that he possessed the fundamental speed necessary for championship success, making his inconsistencies even more frustrating for supporters.

The emergence of other contenders during this period added depth and unpredictability to championship competition. Cal Crutchlow's occasional victories with satellite Honda machinery demonstrated that exceptional riding could occasionally overcome equipment disadvantages, while riders like Alex Rins and Joan Mir began showing the potential that would define their later careers.

Jack Miller's development with Ducati provided another compelling narrative thread throughout this period. The Australian's transition from satellite Ducati rider to factory team member showcased the opportunities available to riders who could adapt to the Italian manufacturer's unique requirements. His aggressive riding style and willingness to push limits in all conditions made him a consistent race winner and championship contender.

The depth of competition during the Marquez era was remarkable compared to previous periods of single-rider dominance. While Marquez's statistical superiority was clear, races were often decided by small margins, with multiple riders capable of victory on any given weekend. This competitive depth made Marquez's dominance even more impressive while ensuring that races remained compelling throughout the era.

The influence of different manufacturers' technological approaches created fascinating strategic elements during this period. Ducati's straight-line speed advantages, Yamaha's cornering capabilities, and Honda's overall balance meant that circuit characteristics often predetermined competitive hierarchies for specific weekends. This technical diversity forced riders to excel across multiple skill areas to achieve championship success.

The psychological pressure of competing against Marquez affected different riders in various ways. Some, like Dovizioso, seemed to thrive under the pressure of championship competition, raising their performance levels to match the challenge. Others struggled with the mental demands of sustaining title campaigns against such a formidable opponent, highlighting the psychological aspects of premier class competition.

Race strategies evolved significantly during this period as teams adapted to the competitive dynamics created by Marquez's presence. Traditional approaches to tire management and race pace became inadequate against a rider who could attack throughout entire race distances while maintaining competitive laptimes. This evolution forced all competitors to reconsider fundamental assumptions about race strategy and risk management.

The media narratives surrounding these championship battles created compelling storylines that attracted audiences beyond traditional motorcycle racing fans. The contrast between different riding styles, nationalities, and manufacturer philosophies provided rich material for broadcasters and journalists seeking to explain the sport's appeal to broader audiences.

By 2019, the competitive ecosystem surrounding Marquez had evolved into one of the most compelling and challenging environments in motorsport history. While his dominance remained clear, the quality of opposition and the narrow margins separating victory from defeat created racing that was both statistically lopsided and competitively thrilling—a combination that defined this era's unique character.

SEVEN

Engineering Excellence

The heart of every MotoGP machine beats with mechanical precision that represents the absolute pinnacle of motorcycle engineering. The evolution of engine technology in the premier class tells the story of relentless pursuit of power, efficiency, and reliability under the most demanding conditions imaginable. From the early four-stroke 990cc era that began in 2002 to today's sophisticated 1000cc powerplants, each developmental phase has pushed the boundaries of what's possible within the confines of technical regulations.

When the four-stroke era began in 2002, manufacturers faced an unprecedented challenge. After decades of two-stroke supremacy, engineering teams had to essentially start from scratch, developing entirely new powerplants that could match the performance characteristics riders had grown accustomed to. The initial 990cc formula allowed for tremendous creativity in engine configuration, leading to a fascinating diversity of approaches. Honda's RC211V featured a revolutionary 75.5-degree V5 configuration, a layout that had never been seen in Grand Prix racing. This unconventional design allowed Honda's engineers to optimize mass centralization while achieving exceptional power delivery characteristics.

Yamaha took a more conservative but equally effective approach with their M1, utilizing a traditional inline-four configuration that leveraged their decades of experience in both racing and road bike development. The Japanese manufacturer's decision to maintain familiar architecture paid dividends in reliability and development efficiency, allowing them to focus resources on advanced electronics and chassis refinement. Meanwhile, Ducati entered the fray with their distinctive L4 configuration, essentially two V-twin engines joined at the crankcase, creating a unique 90-degree layout that would become their signature approach for the next two decades.

The power output of these early four-stroke machines was staggering, with top-tier bikes producing well over 240 horsepower at the rear wheel. This represented a significant increase over the final generation of 500cc two-strokes, but the delivery characteristics were markedly different. Where two-stroke engines provided explosive, almost violent power delivery that required exceptional rider skill to manage, the four-stroke units offered more progressive power curves that, paradoxically, allowed riders to access even higher levels of performance.

Fuel injection systems became increasingly sophisticated during this period, with manufacturers developing proprietary engine management systems that could adjust fuel delivery, ignition timing, and throttle response in real-time based on dozens of sensor inputs. These early electronic systems laid the groundwork for the advanced rider aids that would follow, but their primary function was optimizing combustion efficiency and power delivery across the rev range.

The 800cc era from 2007 to 2011 represented a fascinating experiment in power limitation that ultimately proved the law of unintended consequences. Designed to reduce top speeds and improve safety, the smaller displacement regulations instead led to a new arms race in engine development. Manufacturers compensated for reduced displacement by pushing rev limits ever higher, with engines routinely spinning to 18,000 rpm and beyond. Honda's

RC212V became legendary for its screaming exhaust note and ability to maintain power output well into the stratosphere of engine speeds.

This period saw remarkable innovations in internal engine components, with manufacturers developing exotic materials and manufacturing techniques to withstand the enormous stresses of high-rev operation. Pneumatic valve systems, borrowed from Formula One technology, became standard across the grid, allowing engines to maintain valve control at previously impossible engine speeds. Titanium became commonplace for connecting rods and valves, while advanced ceramics found applications in bearing surfaces and combustion chamber components.

The return to 1000cc in 2012 marked another evolutionary leap, but with crucial differences from the original 990cc era. Fuel capacity limitations became a critical factor, restricting bikes to just 20 liters for race distance. This constraint forced manufacturers to focus intensively on combustion efficiency, leading to remarkable advances in fuel injection technology, combustion chamber design, and engine mapping strategies. Modern MotoGP engines achieve thermal efficiency figures that rival the most advanced automotive powerplants, extracting maximum energy from every drop of fuel.

Contemporary powerplants represent the culmination of decades of development, producing over 280 horsepower while maintaining remarkable reliability over race distance. Advanced materials science has enabled the creation of engine components that were unimaginable just a generation ago. Carbon fiber is now used for structural engine components, while advanced coatings and surface treatments extend component life under extreme operating conditions.

The seamless shift gearbox represents perhaps the most significant recent innovation in MotoGP powerplant technology. Developed initially by Honda and subsequently adopted across the grid, these transmission systems eliminate the traditional power interruption during gear changes, providing seamless acceleration that was previously impossible. The technology requires incredibly precise synchronization between clutch operation, gear selection, and engine management, representing a triumph of electronic control systems working in perfect harmony.

***

The chassis of a modern MotoGP motorcycle represents a masterpiece of engineering optimization, where every gram of weight, every degree of steering geometry, and every millimeter of suspension travel has been carefully calculated to extract maximum performance from the available grip. The evolution from traditional steel trellis frames to today's sophisticated carbon fiber and aluminum constructions tells a story of relentless pursuit of the perfect balance between rigidity and flexibility, strength and weight.

The foundation of any motorcycle's handling characteristics lies in its frame design, and MotoGP has seen a fascinating evolution in structural philosophy over the past two decades. The early four-stroke era saw manufacturers experimenting with various approaches to chassis construction, from Honda's revolutionary carbon fiber frame on the RC211V to Yamaha's more conventional aluminum beam design. Each approach represented different philosophies about how a motorcycle frame should interact with the suspension components and transmit forces from the road surface to the rider.

Honda's carbon fiber experiment represented one of the most ambitious engineering projects in MotoGP history. The RC211V's frame was constructed from layers of carbon fiber laid up in complex orientations to provide precisely tuned stiffness characteristics in different directions. The material allowed Honda's engineers to create a structure that was incredibly light while providing tailored flex characteristics that could be adjusted through changes in fiber orientation and layup patterns. However, the technology proved difficult to develop and costly to produce, leading Honda to eventually return to more conventional aluminum construction.

Yamaha's aluminum twin-beam frame design became the template that most manufacturers would eventually follow, proving that conventional materials could achieve exceptional performance when properly engineered. The Yamaha frame utilized precision-cast aluminum sections welded together to create a structure that provided excellent torsional rigidity while allowing controlled longitudinal flex. This design philosophy recognized that some frame flexibility could actually improve tire contact and rider feedback, contrary to the traditional racing belief that maximum rigidity was always preferable.

Ducati's approach remained distinctively different, maintaining their steel trellis frame construction even as other manufacturers moved to aluminum. The Italian manufacturer's frame design utilized high-strength steel tubes arranged in a complex geometric pattern that distributed loads efficiently throughout the structure. While heavier than aluminum alternatives, the steel construction provided exceptional durability and allowed for easier modification of geometry and stiffness characteristics during development.

The role of chassis geometry in determining handling characteristics cannot be overstated. Modern MotoGP bikes feature incredibly steep steering head angles, typically around 23 degrees, which provide the quick steering response necessary for the tight confines of many Grand Prix circuits. However, this aggressive geometry must be carefully balanced with wheelbase length and swingarm design to maintain stability at the enormous speeds these machines achieve on straight sections.

Suspension technology has undergone equally dramatic evolution, progressing from relatively simple telescopic forks and twin-shock rear systems to sophisticated electronic systems that can adjust damping characteristics hundreds of times per second. The adoption of Öhlins as the spec suspension supplier in recent years has leveled the playing field somewhat, but manufacturers continue to develop proprietary mounting systems and geometry configurations that optimize the performance of these advanced components.

Modern front suspension systems utilize inverted telescopic forks with complex multi-stage compression and rebound damping circuits. These systems can provide dramatically different characteristics for different portions of the suspension travel, allowing for plush compliance over small bumps while maintaining firm control during heavy braking or aggressive cornering. The integration of electronic control systems allows these characteristics to be adjusted in real-time based on sensor inputs measuring everything from lean angle to throttle position.

Rear suspension design has similarly advanced, with most manufacturers now utilizing progressive linkage systems that provide rising rate characteristics as the suspension compresses. These designs allow for optimal tire contact patch management throughout the suspension travel while maintaining the ground clearance necessary for extreme lean angles. The precise tuning of these linkage ratios represents one of the most critical aspects of chassis setup, affecting everything from traction under acceleration to stability under braking.

The development of carbon fiber wheels represents another significant advancement in chassis technology. These components, now used by several manufacturers, offer dramatic weight reduction compared to traditional aluminum wheels while providing improved stiffness characteristics that enhance feedback and precision. The reduced unsprung weight improves suspension performance and allows for quicker direction changes, though the higher cost and reduced durability compared to aluminum wheels continues to limit their adoption.

Electronic suspension systems have begun to appear on MotoGP machines, allowing real-time adjustment of damping characteristics based on riding conditions. These systems use accelerometers and gyroscopes to detect the motorcycle's movement and automatically adjust suspension settings for optimal performance. While still in their infancy, these technologies point toward a future where suspension systems will continuously optimize themselves for changing track conditions and riding situations.

Weight distribution represents another critical aspect of chassis design, with manufacturers working to achieve optimal balance between front and rear wheel loading. Modern MotoGP bikes typically carry approximately

45-50% of their weight over the front wheel, a distribution that provides optimal braking performance while maintaining sufficient rear wheel loading for traction under acceleration. Achieving this balance requires careful consideration of component placement, fuel tank design, and rider positioning.

The integration of aerodynamic elements has introduced new challenges for chassis designers, as the downforce generated by wings and fairings creates additional loads that must be accommodated by the frame structure. These forces can be substantial at high speeds, requiring reinforcement of mounting points and careful consideration of how aerodynamic loads affect handling characteristics at different speeds and lean angles.

***

The transformation of safety standards in MotoGP represents one of the most significant evolutionary changes in the sport's history, driven by both tragic accidents and remarkable technological innovations that have fundamentally altered the risk profile of motorcycle racing at the highest level. From the basic leather suits and open-face helmets of the sport's early decades to today's sophisticated protection systems and emergency response protocols, the commitment to rider safety has become as important as the pursuit of speed itself.

The catalyst for modern safety innovation came through hard-learned lessons, with each significant accident providing impetus for new protective technologies and procedural improvements. The 1993 accident that ended Wayne Rainey's career marked a turning point in the sport's approach to safety, highlighting the need for comprehensive spinal protection and immediate medical response capabilities. This tragedy, along with other serious incidents throughout the 1990s, led to fundamental changes in both personal protective equipment and circuit safety infrastructure.

Modern protective equipment represents a quantum leap from the leather suits and basic helmets that were standard just two decades ago. Today's racing suits incorporate multiple layers of advanced materials, with outer shells of kangaroo leather providing abrasion resistance superior to traditional cowhide, while internal layers of aramid fibers and high-tech polymers absorb impact energy and resist tearing under extreme conditions. The development of these materials often pushes the boundaries of textile engineering, with manufacturers creating fabrics specifically designed to protect human skin during slides at speeds exceeding 200 miles per hour.

The integration of armor inserts within racing suits has evolved from simple foam padding to sophisticated composite structures that distribute impact forces across larger areas of the body. Modern back protectors utilize aerospace-grade materials and advanced geometric designs that provide exceptional impact resistance while maintaining the flexibility necessary for riding. These systems often incorporate multiple layers of different materials, each optimized for specific types of impacts, from the sharp forces encountered in initial contact with barriers to the sustained abrasion of long slides across track surfaces.

Helmet technology has similarly advanced dramatically, with modern designs incorporating complex multi-density foam liners, advanced shell materials, and sophisticated ventilation systems. The adoption of carbon fiber for helmet shells has provided exceptional strength-to-weight ratios, while advanced foam technologies can absorb impact energy more effectively than traditional materials. The development of the HANS (Head and Neck Support) device, originally created for automotive racing, has been adapted for motorcycle use, providing crucial protection against basilar skull fractures and other severe head and neck injuries.

Airbag systems represent perhaps the most revolutionary safety innovation in recent MotoGP history. These systems, pioneered by companies like Alpinestars and Dainese, utilize sophisticated sensors to detect when a rider has lost control of their motorcycle, instantly inflating protective chambers around the chest, back, and shoulders. The technology requires incredibly precise timing and sensor accuracy, as the system must distinguish between normal riding dynamics and genuine accident scenarios within milliseconds. The protective effect of these systems has been dramatic, with numerous documented cases of riders walking away from accidents that would previously have resulted in serious chest or spinal injuries.

The sensors used in airbag systems represent remarkable achievements in miniaturization and precision engineering. Modern systems incorporate GPS receivers, accelerometers, gyroscopes, and pressure sensors, all processing data at frequencies measured in thousands of cycles per second. Machine learning algorithms analyze this data stream to recognize accident patterns and trigger inflation at precisely the right moment. The inflation systems themselves utilize small explosive charges similar to automotive airbag technology, but adapted to provide protection for the specific injury patterns common in motorcycle accidents.

Circuit safety infrastructure has undergone equally dramatic improvements, with every aspect of track design and safety equipment subject to continuous refinement. The development of advanced barrier systems has virtually eliminated the rigid concrete walls that once lined many circuits, replacing them with energy-absorbing structures that dramatically reduce the forces transmitted to riders during impacts. Modern tire barriers incorporate sophisticated backing systems that allow controlled deformation while preventing rider trajectories that could lead to secondary impacts.

The introduction of air fence systems at many circuits has provided another layer of protection, particularly in areas where traditional barriers might create hazardous rider trajectories. These inflatable barriers can absorb enormous impact energies while allowing riders to slide safely along their surfaces, significantly reducing the severity of injuries during high-speed accidents. The positioning and configuration of these systems requires detailed analysis of accident patterns and rider trajectories, with safety experts using computer modeling to optimize placement for maximum protective effect.

Gravel trap design has evolved from simple aggregate surfaces to carefully engineered systems that provide controlled deceleration while minimizing the risk of rider tumbling or catching. Modern gravel specifications consider particle size, depth, and moisture content, with some facilities incorporating multiple zones of different materials to provide optimal deceleration characteristics for different impact scenarios. The maintenance of these systems has become a specialized discipline, with track operators employing dedicated equipment and procedures to ensure consistent protective performance.

Emergency response protocols represent another critical aspect of modern safety systems, with every MotoGP event featuring comprehensive medical support that rivals major hospital trauma centers. The famous "Clinica Mobile" medical unit, led by Dr. Claudio Costa for many years, established the template for trackside medical care that has been adopted throughout motorsport. Modern medical facilities at MotoGP events include fully equipped trauma centers, helicopter landing pads, and direct communication links with nearby hospitals equipped to handle specialized racing injuries.

The medical helicopter protocol, where helicopters are positioned at strategic locations around each circuit with engines running during all on-track sessions, ensures that seriously injured riders can be transported to appropriate medical facilities within minutes of an accident. This rapid response capability has proven crucial in numerous incidents, with the time saved in initial treatment often being the difference between full recovery and permanent disability.

Communication systems linking corner marshals, race control, and medical facilities have become increasingly sophisticated, utilizing digital radio networks and real-time data transmission to coordinate response efforts. Modern systems can automatically alert medical teams when sensors detect serious accidents, often before human observers have fully assessed the situation. This integration of technology and human expertise has created response systems that are both faster and more comprehensive than ever before in the sport's history.

EIGHT

Legendary Circuits

The great European circuits of MotoGP represent more than mere racing venues; they are sacred ground where the sport's most cherished traditions have been forged through decades of extraordinary competition. These legendary tracks have witnessed the rise and fall of champions, the evolution of motorcycle technology, and the passionate devotion of fans who consider these places pilgrimage destinations. Each circuit possesses a unique character that challenges riders in different ways while creating the dramatic moments that define motorcycle racing's rich heritage.

Mugello stands as perhaps the most beloved circuit on the modern MotoGP calendar, a temple of speed nestled in the rolling hills of Tuscany that embodies everything romantic about motorcycle racing. Built in 1974 as Ferrari's private test facility, the circuit was designed with an almost artistic sensibility that creates natural amphitheaters for spectators while challenging riders with a perfect blend of high-speed sections and technical corners. The track's 5.245-kilometer layout flows through the landscape with organic grace, utilizing elevation changes and natural contours to create one of the most visually spectacular racing venues in the world.

The circuit's signature characteristic is its incredible speed, with modern MotoGP bikes reaching velocities that exceed 350 kilometers per hour on the main straight. This creates a unique challenge for riders, who must demonstrate exceptional bravery and precision when following close behind competitors at these extreme velocities. The slipstream effect becomes crucial for overtaking opportunities, leading to dramatic late-braking maneuvers into the first chicane that have produced some of the most memorable moments in MotoGP history.

Mugello's fan culture represents something truly special in motorsport, with the Tuscan crowd bringing an almost operatic passion to their support of racing. The sea of yellow that fills the hillsides during Valentino Rossi's dominance created an atmosphere unlike anything else in motorsport, with tens of thousands of fans creating a wall of sound that could be heard over the screaming engines. This passionate support extends beyond individual riders to embrace the sport itself, with knowledgeable spectators appreciating technical excellence and dramatic racing regardless of nationality.

The track's most challenging section is the sequence from turn six through turn nine, where riders must navigate a series of quick direction changes while managing the transition from full acceleration back to heavy braking. This section has become a crucial overtaking zone in recent years, as improved electronics and aerodynamics have made passing more difficult on the main straight. The technical demands of this section often separate the truly elite riders from the rest of the field, requiring absolute precision at speeds that leave no margin for error.

Assen holds the distinction of being the only circuit to have held a motorcycle Grand Prix in every year since the World Championship began in 1949, earning it the reverence due to such historical significance. Known as "The Cathedral" of motorcycle racing, the Dutch circuit represents the spiritual home of the sport, where tradition and modernity coexist in perfect harmony. The current 4.542-kilometer layout has evolved significantly from its original public road configuration, but it maintains the character and challenge that have made it legendary among riders and fans alike.

The circuit's unique character derives from its combination of flowing corners that allow riders to maintain rhythm and momentum while including technical sections that demand absolute precision. The famous chicane sequence, introduced in 2006 to improve safety, initially drew criticism from traditionalists but has proven to create excellent overtaking opportunities while maintaining the track's essential character. The modification demonstrates how classic circuits can evolve while preserving their fundamental appeal.

Weather plays a crucial role at Assen, with the unpredictable Dutch climate often creating mixed conditions that add an extra dimension to racing. The track's location near the North Sea makes it susceptible to sudden weather changes, and many of the circuit's most memorable races have been decided by tactical decisions regarding tire choice and race strategy during changing conditions. These variables create opportunities for inspired rides and strategic brilliance that have become part of Assen's legendary status.

The Dutch Grand Prix crowd brings their own unique character to the event, with a knowledgeable and appreciative audience that understands the technical aspects of motorcycle racing. The famous orange-clad supporters create a festival atmosphere throughout race weekend, but their appreciation extends beyond nationalism to embrace excellent racing from any competitor. This combination of passionate support and technical appreciation creates an atmosphere that riders consistently cite as among their favorite venues.

Silverstone represents the third pillar of European motorcycle racing tradition, though its history with MotoGP has been more complex than its counterparts. The British circuit's aviation heritage, built on a World War II airfield, gives it a unique character among racing venues, with wide open spaces and fast, sweeping corners that reward brave riding and precise machine setup. The current 5.891-kilometer configuration, introduced in 2010, has created one of the most challenging and rewarding tracks on the modern calendar.

The circuit's signature characteristics include the high-speed Maggotts and Becketts complex, where modern MotoGP bikes can maintain incredible speeds through a series of direction changes that test both machine and rider to their absolute limits. This section has become one of the most spectacular viewing areas in motorsport, with bikes leaned over at extreme angles while maintaining speeds that would be considered dangerous on most public roads. The sight of MotoGP machines threading through this sequence represents motorcycle racing at its most athletic and artistic.

British racing fans bring their own special character to MotoGP events, with a deep appreciation for technical excellence and sporting achievement that transcends national boundaries. The Silverstone crowd's knowledge and enthusiasm create an atmosphere of genuine sporting appreciation, where excellent riding from any competitor receives recognition and applause. This tradition of fair play and sporting appreciation reflects the best aspects of British motorsport culture.

The weather factor at Silverstone adds another dimension to racing, with the unpredictable British climate often creating challenging conditions that test every aspect of team preparation and rider skill. The track's exposure to prevailing winds can create significantly different conditions in various sections, requiring careful attention to setup and strategy. Many of Silverstone's most memorable races have been decided by tactical brilliance during changing weather conditions.

***

The expansion of MotoGP beyond its European heartland has created some of the most spectacular and challenging racing venues in the sport's history. These exotic destinations have brought motorcycle racing to new audiences while creating unique challenges for riders and teams that must adapt to different climates, time zones, and cultural environments. Each of these circuits possesses characteristics that make them special in the MotoGP calendar, contributing to the sport's global appeal while testing competitors in ways that European tracks cannot match.

Phillip Island stands as perhaps the most naturally beautiful racing circuit in the world, perched on the southern coast of Australia where the Southern Ocean meets Bass Strait. The track's dramatic setting, with elevation changes that reveal stunning ocean vistas, creates a racing environment that is both visually spectacular and technically demanding. Built in 1956, the current 4.448-kilometer layout has evolved to become one of the most challenging and rewarding circuits for motorcycle racing, where natural terrain features create corners that cannot be replicated anywhere else.

The circuit's most distinctive characteristic is its incredible speed and flow, with sweeping corners that allow modern MotoGP bikes to maintain extraordinary velocities throughout most of the lap. The combination of constant radius corners and flowing elevation changes creates a rhythm that rewards riders who can find the perfect balance between aggression and precision. Phillip Island has produced some of the closest and most exciting races in MotoGP history, with the track's characteristics often creating large groups of riders separated by mere seconds.

Wind conditions at Phillip Island add another layer of complexity that makes the venue unique among MotoGP circuits. The track's exposed location means that wind direction and strength can vary significantly throughout a race weekend, affecting aerodynamics and bike stability in ways that require constant adaptation. The famous "Phillip Island wind" has influenced the outcome of many races, with riders and teams that best adapt to changing conditions often finding unexpected advantages.

The Australian crowd brings a relaxed but knowledgeable enthusiasm to MotoGP that reflects the country's sporting culture. The accessibility of the circuit allows fans to move freely between viewing areas, creating a festival atmosphere that combines serious racing appreciation with the laid-back Australian lifestyle. The remoteness of the venue, located over 100 kilometers from Melbourne, creates a sense of pilgrimage for international visitors while serving the local community that has embraced motorcycle racing as part of their identity.

Wildlife encounters represent an unusual but genuine concern at Phillip Island, with the surrounding natural environment home to various species that occasionally venture onto the racing circuit. The track's location within a nature reserve means that environmental considerations play a role in event planning, and the presence of little penguins and other native animals adds to the venue's unique character. These considerations demonstrate how MotoGP has learned to coexist with natural environments in ways that benefit both sport and conservation.

The Sepang International Circuit in Malaysia represents the pinnacle of purpose-built racing facility design, created specifically to bring Formula One and motorcycle racing to Southeast Asia. Opened in 1999, the Hermann Tilke-designed circuit incorporates the latest thinking in track design while adapting to the unique challenges of tropical climate racing. The 5.543-kilometer layout features a combination of long straights and technical sections that create multiple overtaking opportunities while testing every aspect of motorcycle and rider performance.

Temperature and humidity at Sepang create physical challenges for riders that are unmatched anywhere else on the MotoGP calendar. Track surface temperatures can exceed 60 degrees Celsius, while ambient temperatures combined with extreme humidity create conditions that push human endurance to its limits. Riders must undergo specialized preparation for the Malaysian Grand Prix, with hydration strategies and physical conditioning programs designed specifically for tropical racing conditions.

The circuit's design cleverly incorporates multiple viewing areas connected by air-conditioned walkways, allowing spectators to enjoy comfortable viewing conditions despite the challenging climate. The facility represents the modern approach to circuit design, where spectator comfort and media facilities receive equal consideration with racing characteristics. The result is a venue that can host world-class motorsport while providing a comfortable experience for attendees from around the globe.

Malaysian fans have embraced MotoGP with enthusiasm that reflects the country's motorcycle culture, where two-wheeled transportation is an integral part of daily life. The crowd's appreciation for technical excellence and close racing creates an atmosphere that riders consistently praise, with passionate but knowledgeable spectators who understand the nuances of motorcycle racing. The diversity of the Malaysian audience, representing the country's multicultural society, creates a truly international atmosphere that embodies MotoGP's global character.

The afternoon thunderstorms that frequently occur at Sepang add an element of strategic complexity that can completely alter race outcomes. These weather patterns are so predictable that they have become part of the circuit's character, with teams developing specific strategies for races that may encounter changing conditions. The ability to read weather conditions and make tactical decisions regarding tire choice and race pace has become crucial for success at this venue.

Qatar represents MotoGP's most ambitious venture into new territory, bringing premier motorcycle racing to the Middle East with a purpose-built facility that showcases the region's commitment to international motorsport. The Losail International Circuit, located just outside Doha, was specifically designed for motorcycle racing, making it unique among Middle Eastern motorsport venues. The track's 5.380-kilometer layout incorporates flowing corners and long straights that create excellent racing while adapting to the desert environment and extreme climate.

The decision to hold the Qatar Grand Prix as a night race represents one of the most innovative developments in modern MotoGP, creating a spectacular visual experience while solving the practical problem of extreme daytime temperatures. The circuit's lighting system, utilizing over 3,000 individual fixtures, creates daylight-like conditions that allow normal racing while producing dramatic visual effects for television audiences worldwide. This innovation has influenced circuit design and event scheduling across motorsport.

The desert environment creates unique challenges for both motorcycles and riders, with sand infiltration requiring special attention to air filtration systems and mechanical components. The extreme temperature variations between day and night practice sessions create setup challenges that require careful attention to tire and suspension characteristics. Teams have developed specialized procedures for desert racing that have become standard practice for the Qatar event.

Cultural considerations play a significant role in the Qatar Grand Prix, with the event serving as a showcase for Middle Eastern hospitality and international cooperation. The race has become an important diplomatic and cultural event, bringing together people from across the region and around the world in celebration of motorsport. This cultural dimension adds meaning beyond pure racing, demonstrating sport's power to bridge cultural differences and create common ground for international understanding.

***

The newest generation of MotoGP circuits represents the culmination of decades of experience in track design, incorporating advanced technology and contemporary understanding of both racing requirements and spectator experience. These modern venues demonstrate how purpose-built facilities can honor the sport's traditions while embracing innovations that enhance safety, competition, and entertainment value. The Circuit of the Americas in Austin, Texas, stands as the flagship example of this new generation, while planned venues around the world promise to continue this evolution.

The Circuit of the Americas (COTA) embodies the American approach to motorsport venue design, where spectacle and technical excellence combine to create an unforgettable experience. Opened in 2012, the Hermann Tilke-designed circuit incorporates signature elements borrowed from legendary tracks around the world while adding uniquely American characteristics that reflect the country's motorsport culture. The 5.513-kilometer layout includes dramatic elevation changes, technical corner sequences, and long straights that create multiple overtaking opportunities throughout each lap.

COTA's most recognizable feature is the dramatic first sector, beginning with a steep uphill climb to turn one that provides both a technical challenge for riders and spectacular viewing for spectators. This section, inspired by legendary corners from circuits around the world, creates an immediate test of commitment and skill that often determines grid positions and race outcomes. The elevation change of over 40 meters across the lap creates unique challenges for motorcycle setup, requiring careful attention to weight distribution and suspension characteristics.

The circuit's design philosophy emphasizes creating multiple viable racing lines through each corner complex, encouraging close racing and overtaking opportunities. This approach represents modern understanding of what creates exciting motorcycle racing, moving beyond the single-groove tracks of earlier eras to provide options for different racing styles and strategies. The result is consistently close racing that showcases the skills of MotoGP riders while creating entertainment value for spectators and television audiences.

American fans have embraced MotoGP with enthusiasm that reflects the country's diverse motorsport culture, bringing elements from NASCAR, IndyCar, and drag racing to create a uniquely American Grand Prix atmosphere. The Austin crowd combines knowledgeable appreciation for technical excellence with a party atmosphere that makes the United States Grand Prix one of the most enjoyable events on the calendar. The integration of music and entertainment with racing creates a festival environment that attracts fans beyond traditional motorcycle racing enthusiasts.

The economic impact of COTA demonstrates the potential for modern MotoGP venues to serve as economic development catalysts for their regions. The circuit's construction and ongoing operations have created thousands of jobs while attracting international visitors who contribute significantly to the local economy. This model of motorsport-driven economic development has influenced planning for new circuits around the world, demonstrating how racing venues can serve broader community interests.

Technology integration at COTA represents the state of the art in modern circuit design, with comprehensive data collection systems that monitor everything from track surface conditions to spectator flow patterns. These systems provide real-time information to race organizers, teams, and broadcasters while creating archives of performance data that contribute to ongoing safety and competition improvements. The facility serves as a testing ground for new technologies that may eventually be implemented at other venues.

Future venue development in MotoGP reflects the sport's continued global expansion, with new markets in Asia, the Americas, and potentially Africa being explored for circuit development. These projects must balance the requirements for world-class racing facilities with local economic conditions and cultural considerations, creating challenges that require innovative approaches to circuit design and financing. The success of venues like COTA provides a template for these developments while demonstrating the potential benefits for host communities.

The evolution toward sustainable circuit design represents an important trend in new venue development, with environmental considerations playing an increasingly important role in project planning. Modern circuits incorporate renewable energy systems, water conservation technologies, and sustainable construction materials that reduce their environmental impact while maintaining world-class racing standards. This approach reflects growing awareness of environmental responsibility within motorsport while demonstrating how racing venues can serve as showcases for sustainable technology.

The integration of digital technology and social media capabilities into modern circuit design creates new opportunities for fan engagement and event promotion. New venues incorporate comprehensive WiFi networks, digital signage systems, and interactive technologies that allow spectators to access real-time information about racing while sharing their experiences through social media platforms. These capabilities enhance the spectator experience while extending the reach of events to global audiences.

Safety considerations in modern circuit design reflect decades of accumulated knowledge about motorcycle racing accidents and injury prevention. New venues incorporate the latest barrier technologies, run-off area designs, and emergency response capabilities from the initial design phase rather than as subsequent modifications. This comprehensive approach to safety planning creates racing environments that allow riders to compete at their maximum potential while minimizing the risks inherent in premier-level motorcycle racing.

The financial models for new circuit development have evolved to reflect the realities of modern motorsport economics, with facilities designed to host multiple event types throughout the year rather than depending solely on MotoGP for economic viability. These multi-use designs allow venues to serve their communities through driving experiences, corporate events, and other motorsport activities while maintaining the specialized requirements necessary for world championship racing.

NINE

The Human Side of MotoGP

The relationship between MotoGP and its fans transcends typical sporting fandom, evolving into a global community united by shared passion for the artistry and athleticism of motorcycle racing at its highest level. This cultural phenomenon reaches its most visible expression in the legendary "Yellow Army" that followed Valentino Rossi throughout his career, but extends far beyond any single rider to encompass a worldwide brotherhood and sisterhood of enthusiasts who find meaning, identity, and community through their connection to the sport.

The Yellow Army represents the most recognizable fan movement in modern motorsport, a sea of yellow-clad supporters whose devotion to Valentino Rossi transformed MotoGP events into celebrations that transcended racing itself. Beginning in the late 1990s and reaching peak intensity during Rossi's championship years, this movement demonstrated the power of charismatic personality to create emotional connections that last across generations. The visual impact of tens of thousands of fans wearing yellow and displaying creative banners created an atmosphere that elevated races from sporting events to cultural celebrations.

What made the Yellow Army phenomenon particularly remarkable was its international character, with passionate Rossi supporters appearing at circuits around the world regardless of geographic or cultural barriers. Italian fans would travel to Australia, Japanese supporters would appear at European races, and American enthusiasts would make pilgrimages to Mugello, all united by their shared admiration for Rossi's racing artistry and magnetic personality. This global community formed lasting friendships and cultural exchanges that extended far beyond the racing weekends that brought them together.

The creative expression of fan support reached artistic levels during Rossi's era, with supporters creating elaborate banners, costumes, and displays that became integral parts of the MotoGP spectacle. The famous "Valentino's Restaurant" banner that appeared at multiple circuits, complete with mock menu featuring rivals served up in various humorous ways, exemplified the wit and creativity that fan culture brought to racing events. These displays required significant planning, coordination, and expense, demonstrating the depth of commitment that characterizes truly passionate motorsport fandom.

The economic impact of dedicated fan groups like the Yellow Army extends far beyond ticket sales, creating ripple effects throughout the motorsport industry and host communities. Rossi's supporters would book accommodations months in advance, purchase merchandise in quantities that supported entire businesses, and plan elaborate travel itineraries that contributed to tourism in circuit locations. This economic influence gave passionate fan groups significant leverage in the sport's commercial calculations, proving that authentic emotional connections translate into measurable business value.

The decline of the Yellow Army following Rossi's retirement has created space for new forms of fan expression to emerge, with contemporary riders developing their own distinctive supporter communities. Marc Márquez's Spanish fans bring their own passionate intensity to races, while emerging talents like Fabio Quartararo and Francesco Bagnaia are attracting dedicated followings that promise to carry forward the tradition of passionate fan support. These newer communities demonstrate how authentic sporting passion transcends individual personalities to become part of the sport's enduring culture.

The diversity of MotoGP's global fan base reflects the sport's international character, with different regions bringing distinct cultural approaches to their support. Japanese fans demonstrate their appreciation through detailed technical knowledge and respectful appreciation for all competitors, creating atmospheres of focused intensity that reflect their cultural values of respect and precision. Their deep understanding of mechanical aspects and racing strategy often surpasses that found in other markets, leading to more technically informed discussions and appreciation for subtle racing craft.

British fans bring their own unique character to MotoGP events, combining traditional sporting fair play with passionate support for their favorites. The Silverstone crowd's willingness to applaud excellent riding regardless of nationality exemplifies the best aspects of sporting appreciation, while their technical knowledge and understanding of racing history creates informed discussions that enrich the spectator experience. British fan culture's emphasis on underdog support and appreciation for determination often creates special moments for riders who might not receive such recognition elsewhere.

American MotoGP fandom represents a fascinating fusion of different motorsport traditions, with fans bringing elements from NASCAR's party atmosphere, IndyCar's technical appreciation, and motorcycle culture's rebellious independence. The Austin crowd's enthusiasm creates one of the most energetic atmospheres on the calendar, while American fans' embrace of technology and data analysis contributes to increasingly sophisticated online communities where technical discussions reach remarkable depth and insight.

The rise of social media has fundamentally transformed how MotoGP fans connect with each other and the sport itself, creating global communities that exist beyond the physical boundaries of racing events. Online forums, Facebook groups, and Twitter conversations allow fans to share information, debate technical aspects, and maintain year-round engagement with the sport. These digital communities often become as important to fans as attending actual races, providing daily connection to their passion through news, analysis, and social interaction with like-minded enthusiasts.

Fan-generated content has become an increasingly important aspect of MotoGP's cultural landscape, with supporters creating everything from detailed technical analyses to artistic tributes that rival professional productions. YouTube channels dedicated to MotoGP analysis attract hundreds of thousands of subscribers, while fan art, photography, and written content contribute to a rich ecosystem of amateur content creation that supplements official media coverage. This grassroots content creation demonstrates the depth of engagement that MotoGP inspires among its most dedicated followers.

The intergenerational aspect of MotoGP fandom creates unique family traditions where parents pass their passion for the sport to children, creating lifelong bonds built around shared racing experiences. These family traditions often include annual pilgrimages to specific circuits, with multiple generations gathering to witness races together while sharing stories and knowledge accumulated over decades of following the sport. The sight of grandparents explaining racing lines to grandchildren while veteran fans mentor newcomers demonstrates how MotoGP creates communities that transcend age and experience levels.

Women's participation in MotoGP fandom has grown significantly in recent years, challenging traditional assumptions about motorsport demographics while bringing new perspectives to fan communities. Female fans often contribute different forms of engagement, from detailed social media analysis to creative expression through art and writing, enriching the overall fan experience while demonstrating the sport's broadening appeal. Their presence at races and in online communities has helped create more inclusive environments that welcome diverse forms of enthusiasm and engagement.

The phenomenon of "racing family" adoption, where fans develop deep emotional connections to riders and their teams, creates support networks that extend beyond typical entertainment consumption. Fans invest emotionally in riders' careers, celebrating victories with genuine joy while experiencing defeats as personal disappointments. This emotional investment drives the intense loyalty that characterizes MotoGP fandom while creating the passionate atmospheres that make races special for both spectators and competitors.

***

The transformation of MotoGP media coverage represents one of the most dramatic changes in how sporting events are documented, distributed, and consumed by global audiences. From the early days when motorcycle racing received minimal coverage in mainstream publications to today's comprehensive digital ecosystem that provides unprecedented access to every aspect of the sport, this evolution mirrors broader changes in media technology while creating new opportunities for fan engagement and commercial development.

The golden age of motorcycle racing journalism coincided with the sport's rise to prominence in the 1970s and 1980s, when publications like "Motocourse" and specialized racing magazines provided the primary source of detailed coverage for dedicated fans. These publications featured extensive photography, technical analyses, and rider interviews that created the foundational literature of motorcycle racing culture. The annual "Motocourse" volumes became historical documents that preserved the sport's heritage while providing analysis and commentary that shaped how fans understood racing's technical and human dimensions.

Photography played a crucial role in early MotoGP media, with pioneering photojournalists like Henk Keulemans and others capturing images that became iconic representations of racing's beauty and danger. These photographers developed techniques for capturing motorcycles in motion that required remarkable timing and positioning, often placing themselves in dangerous locations to achieve shots that conveyed the speed and intensity of racing. Their work created the visual vocabulary that continues to define how motorcycle racing is perceived and understood by fans worldwide.

Television coverage began modestly in the 1970s but grew to become the dominant medium for MotoGP consumption, fundamentally changing how fans experienced racing while creating new commercial opportunities for the sport. Early broadcast technology limited coverage to highlight packages and delayed transmissions, but advancing technology eventually enabled live worldwide coverage that brought the excitement of racing directly into fans' homes. The development of on-board cameras, slow-motion replay, and sophisticated graphics enhanced television's ability to convey the technical and dramatic aspects of racing.

The introduction of specialized camera angles and innovative filming techniques transformed television coverage from simple documentation to sophisticated entertainment that could capture nuances invisible to trackside spectators. Helicopter-mounted cameras provided sweeping aerial perspectives that revealed racing lines and strategy, while track-side cameras with extreme telephoto lenses could capture facial expressions and body language that conveyed the human drama of competition. These technological capabilities created new forms of storytelling that enhanced understanding while building emotional connections between viewers and competitors.

Commentary teams became crucial intermediaries between the sport and its television audience, with legendary voices like Murray Walker, Julian Ryder, and others developing distinctive styles that educated while entertaining. The best commentators combined technical knowledge with storytelling ability, helping viewers understand complex racing situations while building narrative tension that made even processional races engaging. Their expertise and passion became integral parts of how fans experienced MotoGP, with many supporters developing strong preferences for particular commentary teams.

The digital revolution fundamentally transformed MotoGP media consumption, beginning with websites that provided news and results but eventually encompassing comprehensive streaming services that offer unprecedented access to all aspects of the sport. The official MotoGP VideoPass service represents the current pinnacle of digital sports coverage, providing live streaming of all sessions, historical race archives, and supplementary content that includes technical analysis, rider interviews, and behind-the-scenes documentation that was previously unavailable to fans.

Streaming technology has democratized access to MotoGP content while creating new opportunities for specialized coverage that serves different fan interests. Technical analysis channels can focus on detailed examinations of racing craft and motorcycle technology, while entertainment-focused content can emphasize personality profiles and human-interest stories. This fragmentation allows content creators to serve specific audience niches while contributing to a more comprehensive overall coverage ecosystem.

Social media integration has created real-time interaction opportunities that transform passive media consumption into active community participation. Fans can comment on races as they unfold, share observations and analyses, and engage directly with riders and teams through official social media channels. This immediate interaction capability has created expectations for responsiveness and engagement that have fundamentally changed how MotoGP organizations manage their public communications and fan relationships.

The rise of independent content creators and fan-generated media has supplemented official coverage while providing alternative perspectives and analyses that often rival professional productions in quality and insight. YouTube channels dedicated to MotoGP analysis attract substantial audiences while providing platforms for detailed technical discussions and historical retrospectives that enhance understanding of the sport's complexities. These independent voices contribute to a more diverse media landscape while creating opportunities for passionate fans to share their expertise with broader audiences.

Data analytics and performance metrics have become increasingly important components of modern MotoGP media coverage, with advanced telemetry and statistical analysis providing insights that were previously available only to teams and riders. Modern coverage can display real-time speed, lean angle, and braking force data that allows viewers to understand precisely what riders are accomplishing while providing objective measures of performance that supplement subjective observations about racing craft.

Virtual and augmented reality technologies promise to further revolutionize MotoGP media consumption, with experimental applications already providing immersive experiences that allow fans to experience races from riders' perspectives or explore circuits in three-dimensional detail. These technologies could eventually provide spectator experiences that surpass attending actual races in terms of access and information availability, fundamentally changing relationships between fans and the sport they follow.

The global nature of modern MotoGP media coverage has created challenges and opportunities related to time zones, language barriers, and cultural differences in how racing is perceived and discussed. Streaming services must accommodate viewers in dozens of time zones while providing commentary and analysis in multiple languages, creating production complexities that require sophisticated technical and logistical solutions. However, this global reach has also created opportunities for cultural exchange and shared appreciation that transcends national boundaries.

***

The economic ecosystem surrounding MotoGP represents a complex web of relationships involving manufacturers, sponsors, teams, broadcasters, and circuit operators that generates billions of dollars in economic activity while driving technological innovation throughout the motorcycle industry. Understanding these financial dynamics reveals how sporting passion translates into business value while demonstrating the sport's significance beyond entertainment to include industrial development, technological advancement, and regional economic growth.

Manufacturer investment in MotoGP represents the largest single financial component of the sport's economy, with companies like Honda, Yamaha, Ducati, Suzuki, and KTM spending tens of millions of dollars annually on their racing programs. These expenditures encompass not only direct racing costs but also research and development activities that contribute to road bike development while advancing the technological state of the art across the motorcycle industry. The return on this investment includes brand prestige, technological knowledge, and marketing value that influences consumer purchasing decisions worth billions of dollars annually.

The manufacturer business case for MotoGP participation extends far beyond simple marketing considerations to include strategic advantages in product development and market positioning. Racing success translates directly into consumer confidence in road bike products, with championship victories providing marketing messages that cannot be replicated through traditional advertising. The technical knowledge gained through racing competition accelerates road bike development while providing validation for engineering approaches that influence product design decisions worth hundreds of millions of dollars.

Team economics in MotoGP involve complex relationships between manufacturers, sponsors, and independent organizations that create diverse business models reflecting different strategic approaches. Factory teams operate as direct manufacturer extensions with budgets that can exceed fifty million dollars annually, while satellite teams must balance manufacturer support with independent sponsorship to create viable business models. These different approaches create economic diversity that strengthens the championship while providing opportunities for different types of business participation.

Independent teams play crucial roles in MotoGP's economic ecosystem by providing development opportunities for emerging riders while offering sponsorship platforms for companies that cannot afford factory team partnerships. These organizations often operate on budgets one-tenth the size of factory efforts while maintaining competitive relevance through intelligent resource allocation and specialized expertise. Their success demonstrates how business acumen and technical innovation can overcome financial disadvantages in professional motorsport.

Sponsorship economics in MotoGP have evolved from simple logo placement to comprehensive marketing partnerships that integrate racing participation with broader corporate communication strategies. Major sponsors like Monster Energy, Red Bull, and telecommunications companies use MotoGP platforms to reach global audiences while associating their brands with high-performance technology and international glamour. These partnerships often include activation programs that extend far beyond racing weekends to include consumer experiences and product development collaborations.

Title sponsorship and naming rights represent premium partnership opportunities that provide extensive brand exposure while supporting the fundamental economics of racing operations. Circuit naming rights, championship title sponsorships, and team partnerships provide sustained revenue streams that enable long-term planning and investment in competitive programs. These relationships demonstrate how commercial partnerships can create mutual value while contributing to the sport's overall financial stability.

The television and media rights economy forms another crucial component of MotoGP's financial structure, with broadcast partnerships providing revenue streams that support the championship's global operations while creating the content that attracts sponsor investment. The development of streaming services and digital media platforms has created new revenue opportunities while changing how media value is calculated and distributed throughout the sport's economic ecosystem.

Circuit economics represent a complex balance between hosting fees paid to MotoGP, spectator revenue, and broader economic impact on host regions. Many circuits operate at losses on their MotoGP events while generating broader economic benefits through tourism, international exposure, and associated business development. The success of venues like the Circuit of the Americas demonstrates how racing events can serve as economic development catalysts that justify public and private investment in racing infrastructure.

The technology transfer value of MotoGP extends far beyond direct racing applications to influence development across the broader motorcycle industry and related sectors. Innovations in materials science, electronics, and mechanical engineering that originate in racing programs eventually find applications in road bikes, automotive systems, and other industrial applications. This technology transfer represents significant economic value that is difficult to quantify but contributes substantially to industrial competitiveness and innovation.

Regional economic impact studies consistently demonstrate that MotoGP events generate economic activity far exceeding direct event revenues, with visitors spending money on accommodations, dining, transportation, and tourism activities that benefit broad segments of host communities. These economic impacts justify public sector support for racing venues while demonstrating how motorsport can contribute to regional development strategies and tourism promotion efforts.

The employment impact of MotoGP extends from direct racing positions to supporting industries including specialized manufacturing, logistics, media production, and event management. The sport creates high-value employment opportunities that require specialized skills while contributing to industrial clusters around racing organizations and manufacturing facilities. These employment effects contribute to regional economic development while creating career opportunities in technical fields that might not otherwise exist.

Global merchandise and licensing revenue represents another significant economic component, with official MotoGP products generating substantial revenue streams while creating business opportunities for manufacturers, retailers, and distributors worldwide. The popularity of rider merchandise, team gear, and championship memorabilia creates market opportunities that extend throughout the year rather than being limited to race weekends, providing sustained revenue generation that supports the sport's economic foundation.

Future economic developments in MotoGP will likely focus on digital revenue opportunities, sustainable technology development, and expansion into new geographic markets that can support the sport's continued growth. The integration of electric racing through MotoE represents both a technological challenge and a commercial opportunity that could attract new sponsors and manufacturers while addressing environmental concerns that influence contemporary business decisions. These developments will determine how successfully MotoGP adapts to changing economic conditions while maintaining its position as the premier motorcycle racing championship.

TEN

MotoGP in the Modern Era and Beyond

The landscape of MotoGP underwent a seismic shift in 2020 when Fabio Quartararo became the first French rider to win the premier class championship, breaking a drought that had lasted since the sport's inception. At just 21 years old, Quartararo's victory represented more than a personal triumph; it symbolized the emergence of a new generation capable of challenging the established order that had been dominated by Spanish and Italian riders for over two decades.

Quartararo's path to glory was neither conventional nor predictable. Unlike many of his predecessors who climbed methodically through the smaller classes, the Frenchman's journey to MotoGP was marked by bold career moves and calculated risks. His decision to join the Petronas Yamaha satellite team in 2019, despite offers from other manufacturers, demonstrated a maturity beyond his years. The young rider understood that success in MotoGP required more than raw talent; it demanded the right machinery, the right team environment, and the right timing.

The 2020 championship battle between Quartararo and Joan Mir showcased the evolution of modern MotoGP racing. Gone were the days when a single rider could dominate through superior machinery or exceptional individual brilliance. The new era demanded consistency, adaptability, and mental resilience across a diverse calendar of circuits with varying characteristics. Quartararo's ability to extract performance from the Yamaha M1 across different track configurations highlighted the technical sophistication that modern riders must possess.

Francesco Bagnaia's emergence as Quartararo's primary rival has created one of the most compelling championship battles in recent memory. The Italian's methodical approach to racecraft, combined with Ducati's engineering excellence, has produced a rivalry that transcends national boundaries and manufacturer loyalties. Bagnaia's 2022 championship victory, achieved through a remarkable late-season surge, demonstrated the tactical sophistication that characterizes contemporary MotoGP competition.

The depth of talent in the current MotoGP field extends far beyond the championship protagonists. Riders like Jorge Martin, Enea Bastianini, and Brad Binder have consistently challenged for race victories, creating an unpredictability that keeps fans on the edge of their seats. This competitive parity reflects the sport's evolution toward a more balanced technical landscape, where multiple manufacturers can field championship-capable machinery.

The geographical diversity of the current grid represents MotoGP's successful expansion beyond its traditional European heartland. Australian Jack Miller's aggressive racing style, South African Brad Binder's remarkable adaptability, and Japanese Takaaki Nakagami's technical precision have brought new perspectives to the sport. These riders carry the hopes and dreams of their respective nations, inspiring new generations of motorcyclists in markets that were previously peripheral to the MotoGP narrative.

The current generation has embraced technology and data analysis to an unprecedented degree. Modern MotoGP riders work with sophisticated telemetry systems, biomechanical specialists, and mental performance coaches.

The sport has evolved from one based primarily on natural talent and mechanical sympathy to one where scientific methodology and professional support systems play crucial roles in determining success.

The rise of riders like Pedro Acosta, who made an immediate impact in his rookie MotoGP season, suggests that the talent pipeline remains robust. These young riders bring fresh perspectives and fearless approaches that challenge established conventions. Their comfort with digital technology and social media has also transformed how MotoGP engages with younger audiences, ensuring the sport's continued relevance in an increasingly crowded entertainment landscape.

***

The motorsport industry faces unprecedented pressure to address environmental concerns while maintaining its core appeal of speed, competition, and technological innovation. MotoGP's response to this challenge has been both pragmatic and forward-thinking, acknowledging that the sport's long-term viability depends on its ability to demonstrate environmental responsibility without compromising its fundamental character.

The MotoE World Cup, launched in 2019, represents MotoGP's most visible commitment to sustainable innovation. While initially viewed with skepticism by traditional racing fans, the electric motorcycle championship has evolved into a compelling spectacle that showcases the potential of zero-emission racing. The Energica Ego Corsa machines produce instant torque delivery that creates unique racing dynamics, with riders adapting their techniques to manage energy consumption while maintaining competitive pace.

The technical challenges overcome in developing competitive electric motorcycles have been substantial. Battery technology, energy management systems, and power delivery characteristics required completely new approaches to motorcycle design and racing strategy. The collaboration between Dorna Sports, the FIM, and technology partners like Energica has accelerated developments that benefit both racing and road-going electric motorcycles.

Beyond the MotoE championship, MotoGP has committed to achieving carbon neutrality by 2027. This ambitious goal encompasses all aspects of the championship, from freight transportation and circuit operations to team activities and fan engagement. The development of sustainable synthetic fuels represents a particularly promising avenue for maintaining the spectacle of internal combustion engines while dramatically reducing carbon emissions.

The synthetic fuel initiative, supported by major manufacturers and fuel suppliers, aims to create drop-in replacements for traditional fossil fuels using renewable energy sources. These fuels could maintain the sound, performance characteristics, and mechanical complexity that define traditional motorcycle racing while achieving near-zero net carbon emissions. Early testing has shown promising results, with prototype fuels delivering performance comparable to conventional racing fuels.

Circuit sustainability initiatives have transformed how MotoGP events are planned and executed. Solar power installations, waste reduction programs, and sustainable transportation options have become standard features at many venues. The Circuit de Barcelona-Catalunya's comprehensive sustainability program serves as a model for other venues, demonstrating that environmental responsibility and world-class racing can coexist successfully.

The sport's sustainability efforts extend to encouraging fan behavior and industry practices. Digital ticketing systems, recycling programs, and public transportation partnerships have reduced the environmental impact of race attendance. Simultaneously, MotoGP's influence on motorcycle manufacturers has accelerated the development of cleaner, more efficient road bikes that benefit millions of everyday riders.

Technological innovations driven by sustainability requirements often produce unexpected competitive advantages. Advances in aerodynamics initially developed for efficiency have improved performance. Lightweight materials created for environmental reasons have enhanced handling characteristics. This symbiotic relationship between sustainability and performance suggests that environmental responsibility can drive rather than hinder sporting excellence.

The younger generation of riders and fans has embraced MotoGP's sustainability initiatives with enthusiasm that surpasses that of their older counterparts. For riders like Fabio Quartararo and Francesco Bagnaia, environmental stewardship is not a constraint but an integral part of their professional identity. This generational shift ensures that sustainability will remain a priority as these individuals assume leadership roles within the sport.

***

As MotoGP approaches its 75th anniversary, the championship's influence extends far beyond the boundaries of professional motorcycle racing. The sport has become a catalyst for technological innovation, a platform for international cultural exchange, and an inspiration for millions of enthusiasts worldwide. Understanding this broader impact reveals why MotoGP occupies such a significant place in the global sporting landscape.

The technological legacy of MotoGP permeates virtually every aspect of modern motorcycle design. Anti-lock braking systems, traction control, electronic suspension, and advanced materials that originated in racing laboratories now appear on motorcycles accessible to everyday riders. The sport's relentless pursuit of performance has compressed development timelines that might otherwise span decades into just a few racing seasons.

This technology transfer has democratized access to sophisticated motorcycle capabilities. A rider purchasing a mid-range sport bike today enjoys safety and performance features that were exclusive to factory racing machines just a generation ago. Electronic rider aids that prevent accidents, suspension systems that adapt to changing conditions, and engine management systems that optimize fuel efficiency all trace their heritage to MotoGP development programs.

The sport's influence on motorcycle design philosophy has been equally profound. The emphasis on lightweight construction, aerodynamic efficiency, and ergonomic optimization that characterizes modern motorcycles reflects lessons learned on racing circuits worldwide. Manufacturers invest billions in MotoGP participation partly because the technical knowledge gained provides competitive advantages in showroom motorcycles.

MotoGP's cultural impact transcends engineering and technology. The sport has created global communities of fans who share passion for mechanical excellence, competitive drama, and human achievement under extreme conditions. These communities span geographical, linguistic, and cultural boundaries, united by appreciation for the skill, courage, and dedication that define elite motorcycle racing.

The championship has also served as a vehicle for national pride and international diplomacy. When a rider achieves success in MotoGP, their entire nation celebrates. Valentino Rossi's achievements elevated Italy's profile in international motorsport. Casey Stoner's success inspired Australian interest in motorcycle racing. Marc Márquez's dominance reinforced Spain's position as a global sporting powerhouse. These individual accomplishments create positive associations that benefit their home countries' international relationships and tourism industries.

The educational impact of MotoGP extends into science, technology, engineering, and mathematics fields. Universities worldwide use MotoGP examples to illustrate principles of physics, engineering, and material science. The sport's technological sophistication inspires students to pursue technical careers, contributing to workforce development in advanced manufacturing and engineering sectors.

MotoGP's safety innovations have influenced transportation safety far beyond motorcycles. Helmet technology, protective clothing materials, and impact-absorbing barriers developed for racing have saved lives on public roads worldwide. The sport's comprehensive approach to safety management, including medical response protocols and circuit design standards, has established benchmarks adopted by other motorsports and transportation industries.

The championship's media evolution has paralleled broader changes in global communications. From radio broadcasts to television coverage to digital streaming and social media engagement, MotoGP has consistently adapted to new platforms and technologies. This adaptability has maintained relevance across generations while expanding accessibility to new audiences worldwide.

Perhaps most importantly, MotoGP continues to inspire individuals to pursue excellence in their chosen fields. The dedication, perseverance, and pursuit of marginal gains that characterize successful MotoGP participants provide templates for achievement that transcend motorsport. Young people watching riders overcome adversity, adapt to changing circumstances, and maintain focus under pressure learn valuable life lessons that serve them well beyond racing contexts.

The sport's ability to combine individual achievement with team collaboration offers particularly relevant lessons for modern professional environments. MotoGP riders must balance personal ambition with team objectives, technical expertise with intuitive feel, and risk-taking with calculated decision-making. These skills translate directly to leadership challenges in business, technology, and other competitive fields.

As MotoGP looks toward its second 75 years, the championship's legacy provides a foundation for continued growth and innovation. The sport's history demonstrates that automotive excellence, technological progress, and environmental responsibility can coexist successfully. The riders, manufacturers, and organizations that have shaped MotoGP's development have created not just a sporting spectacle, but a cultural institution that enriches lives worldwide.

The next generation of MotoGP participants inherits a sport that has proved its ability to adapt, innovate, and inspire. Whether the future brings electric powertrains, sustainable fuels, or technologies not yet imagined, MotoGP's core values of speed, skill, and competition will continue to drive human achievement forward. The sport's greatest legacy may be its demonstration that the pursuit of excellence, conducted with integrity and respect for others, can inspire and unite people across all boundaries.

Conclusion

Seventy-five years after the first Grand Prix motorcycle World Championship race, MotoGP stands as more than a sporting competition; it represents humanity's eternal quest to push boundaries, defy limitations, and achieve the seemingly impossible. From the primitive motorcycles that contested those early championships to today's sophisticated machines that blur the line between engineering and art, the sport has maintained an unbroken thread of innovation, courage, and dedication that speaks to the best aspects of human nature.

The journey chronicled in these pages reveals patterns that extend far beyond motorcycle racing. We see how technological innovation drives competitive advantage, how individual brilliance must harmonize with team effort, and how tradition and progress can coexist productively. The riders who have graced MotoGP's circuits over seven decades have demonstrated that excellence requires not just natural talent, but also relentless preparation, mental resilience, and the courage to face danger in pursuit of perfection.

MotoGP's evolution mirrors broader changes in society, technology, and global culture. The sport's expansion from European origins to worldwide phenomenon reflects our increasingly interconnected world. The technological sophistication of modern racing machines parallels advances in computing, materials science, and manufacturing that have transformed daily life. The championship's growing emphasis on sustainability addresses environmental challenges that define our era.

Significantly, MotoGP has consistently demonstrated that competition and cooperation can coexist beneficially. Rival manufacturers share safety innovations. Competing riders support each other in times of need. Nations that might disagree on political matters unite in celebrating sporting achievement. The paddock community that travels the world together has created a unique international culture that transcends traditional boundaries.

The human stories that define MotoGP's legacy remind us that behind every technological marvel and statistical achievement lie individuals who chose to pursue extraordinary goals despite ordinary limitations. Valentino

Rossi's transformation from small-town Italian youth to global superstar. Casey Stoner's decision to retire at the peak of his powers to prioritize family life. Marc Márquez's determination to return from career-threatening injuries. These narratives inspire because they demonstrate that exceptional achievement is possible for those willing to pay the necessary price.

The sport's technological contributions to transportation safety, efficiency, and performance have improved millions of lives worldwide. Every rider who benefits from modern ABS systems, every commuter whose motorcycle consumes less fuel, every accident victim protected by advanced helmet technology has been touched by MotoGP's relentless pursuit of excellence. This practical legacy may ultimately prove more significant than any championship or lap record.

As we look toward MotoGP's future, we see a sport confidently addressing contemporary challenges while preserving its essential character. The development of sustainable racing technologies, the expansion into new markets, and the nurturing of emerging talent demonstrate that MotoGP remains vitally relevant to our changing world. The sport's ability to adapt while maintaining its core identity suggests a bright future for the next 75 years.

The riders who will contest future championships may pilot electric motorcycles powered by renewable energy. They may compete on circuits we cannot yet imagine, using technologies not yet invented. But they will carry forward the same fundamental commitment to speed, skill, and competition that has defined MotoGP since 1949. They will push the boundaries of what is possible, inspire new generations of fans, and continue adding chapters to this remarkable story.

MotoGP's greatest achievement may be its demonstration that the pursuit of excellence, conducted with respect for competitors and concern for safety, can unite rather than divide. In an era of increasing polarization and conflict, the sport offers a model of how competition can elevate all participants rather than diminishing the defeated. The mutual respect between rivals, the support for injured competitors, and the celebration of achievement regardless of nationality or background provide hopeful examples for society at large.

The championship's future will undoubtedly bring changes we cannot fully anticipate. New technologies will emerge, new challenges will arise, and new heroes will capture our imagination. But the fundamental appeal of

MotoGP—watching extraordinarily skilled individuals push themselves and their machines to the absolute limit in pursuit of victory—will remain constant. This eternal dance of speed and passion, played out on the world's greatest racing circuits, continues to inspire and amaze because it speaks to something fundamental in human nature.

We are creatures who dream of flying, of moving faster than our limitations should allow, of achieving perfection in imperfect circumstances. MotoGP provides a venue where these dreams become reality, where the impossible becomes routine, where ordinary individuals become legends. As long as humans possess the desire to go faster, to compete, to excel, and to dream, MotoGP will remain relevant and inspiring.

The story of MotoGP is ultimately the story of human potential unleashed through competition, innovation, and dedication. It reminds us that barriers exist to be broken, that limits exist to be transcended, and that the pursuit of excellence can bring out the best in individuals and communities alike. In celebrating MotoGP's remarkable journey, we celebrate our own capacity for achievement, growth, and the endless pursuit of something greater than ourselves.

The checkered flag waves, the engines fall silent, but the inspiration continues. The legacy of 75 years of Grand Prix motorcycle racing lives on in every rider who pushes a little harder, every engineer who seeks a marginal gain, every fan who dreams of achievement. MotoGP's greatest victory is not any single championship, but its demonstration that the human spirit, properly channeled and supported, can achieve extraordinary things.

As the sun sets on one era of MotoGP history, it rises on another filled with unlimited potential. The sport that began with a handful of brave riders on primitive machines has evolved into a global phenomenon that continues to push the boundaries of technology, human performance, and international cooperation. The best chapters of this remarkable story may yet be written, but the foundation established over 75 years ensures that whatever comes next will be worthy of the legacy that has brought us this far.

MotoGP succeeds because it reminds us who we can become when we refuse to accept limitations as final answers. The sport continues because it speaks to something timeless in human nature—our desire to go faster, to be better, and to achieve something extraordinary. As long as this desire persists, MotoGP will continue to provide a stage where dreams become reality and legends are born. The race continues, the pursuit of perfection never ends, and the story of speed, passion, and human excellence writes itself anew with each passing season.

Appendices

Complete Championship Winners by Year and Class

Premier Class Champions (500cc 1949-2001, MotoGP 2002-Present)

1949-1959: The Foundation Years

- 1949: Leslie Graham (AJS) - 30 points
- 1950: Umberto Masetti (Gilera) - 21 points
- 1951: Geoff Duke (Norton) - 33 points
- 1952: Umberto Masetti (Gilera) - 30 points
- 1953: Geoff Duke (Gilera) - 34 points
- 1954: Geoff Duke (Gilera) - 33 points
- 1955: Geoff Duke (Gilera) - 35 points
- 1956: John Surtees (MV Agusta) - 32 points
- 1957: Libero Liberati (Gilera) - 32 points
- 1958: John Surtees (MV Agusta) - 40 points

- 1959: John Surtees (MV Agusta) - 44 points

1960-1979: The Age of Legends

- 1960: John Surtees (MV Agusta) - 38 points
- 1961: Gary Hocking (MV Agusta) - 42 points
- 1962: Mike Hailwood (MV Agusta) - 52 points
- 1963: Mike Hailwood (MV Agusta) - 54 points
- 1964: Mike Hailwood (MV Agusta) - 56 points
- 1965: Mike Hailwood (MV Agusta) - 54 points
- 1966: Giacomo Agostini (MV Agusta) - 56 points
- 1967: Giacomo Agostini (MV Agusta) - 60 points
- 1968: Giacomo Agostini (MV Agusta) - 60 points
- 1969: Giacomo Agostini (MV Agusta) - 90 points
- 1970: Giacomo Agostini (MV Agusta) - 87 points
- 1971: Giacomo Agostini (MV Agusta) - 96 points
- 1972: Giacomo Agostini (MV Agusta) - 134 points
- 1973: Phil Read (MV Agusta) - 148 points
- 1974: Phil Read (MV Agusta) - 127 points
- 1975: Giacomo Agostini (Yamaha) - 148 points
- 1976: Barry Sheene (Suzuki) - 107 points
- 1977: Barry Sheene (Suzuki) - 103 points
- 1978: Kenny Roberts Sr. (Yamaha) - 108 points
- 1979: Kenny Roberts Sr. (Yamaha) - 113 points

1980-1999: Two-Stroke Mastery

- 1980: Kenny Roberts Sr. (Yamaha) - 116 points
- 1981: Marco Lucchinelli (Suzuki) - 105 points
- 1982: Franco Uncini (Suzuki) - 94 points
- 1983: Freddie Spencer (Honda) - 144 points
- 1984: Eddie Lawson (Yamaha) - 139 points
- 1985: Freddie Spencer (Honda) - 141 points
- 1986: Eddie Lawson (Yamaha) - 139 points
- 1987: Wayne Gardner (Honda) - 153 points
- 1988: Eddie Lawson (Yamaha) - 228 points
- 1989: Eddie Lawson (Honda) - 212 points
- 1990: Wayne Rainey (Yamaha) - 188 points
- 1991: Wayne Rainey (Yamaha) - 201 points
- 1992: Wayne Rainey (Yamaha) - 231 points
- 1993: Kevin Schwantz (Suzuki) - 178 points
- 1994: Mick Doohan (Honda) - 287 points
- 1995: Mick Doohan (Honda) - 306 points
- 1996: Mick Doohan (Honda) - 297 points
- 1997: Mick Doohan (Honda) - 340 points
- 1998: Mick Doohan (Honda) - 260 points
- 1999: Àlex Crivillé (Honda) - 157 points

2000-2024: MotoGP Era

- 2000: Kenny Roberts Jr. (Suzuki) - 163 points
- 2001: Valentino Rossi (Honda) - 325 points

- 2002: Valentino Rossi (Honda) - 355 points
- 2003: Valentino Rossi (Honda) - 357 points
- 2004: Valentino Rossi (Yamaha) - 304 points
- 2005: Valentino Rossi (Yamaha) - 367 points
- 2006: Nicky Hayden (Honda) - 252 points
- 2007: Casey Stoner (Ducati) - 367 points
- 2008: Valentino Rossi (Yamaha) - 373 points
- 2009: Valentino Rossi (Yamaha) - 306 points
- 2010: Jorge Lorenzo (Yamaha) - 383 points
- 2011: Casey Stoner (Honda) - 350 points
- 2012: Jorge Lorenzo (Yamaha) - 350 points
- 2013: Marc Márquez (Honda) - 334 points
- 2014: Marc Márquez (Honda) - 362 points
- 2015: Jorge Lorenzo (Yamaha) - 330 points
- 2016: Marc Márquez (Honda) - 298 points
- 2017: Marc Márquez (Honda) - 298 points
- 2018: Marc Márquez (Honda) - 321 points
- 2019: Marc Márquez (Honda) - 420 points
- 2020: Joan Mir (Suzuki) - 171 points
- 2021: Fabio Quartararo (Yamaha) - 278 points
- 2022: Francesco Bagnaia (Ducati) - 265 points
- 2023: Francesco Bagnaia (Ducati) - 508 points
- 2024: Francesco Bagnaia (Ducati) - 461 points

350cc Class Champions (1949-1982)

Selected Champions:

- 1949: Freddie Frith (Velocette)
- 1966-1972: Giacomo Agostini (MV Agusta) - 7 consecutive titles
- 1974: Giacomo Agostini (Yamaha)
- 1975-1977: Johnny Cecotto (Yamaha)
- 1980: Jon Ekerold (Bimota)
- 1981-1982: Anton Mang (Kawasaki)

250cc Class Champions (1949-2009)

Notable Champions:

- 1962-1965: Jim Redman (Honda) - 4 consecutive titles
- 1971: Phil Read (Yamaha)
- 1981-1984: Anton Mang (Kawasaki) - 4 consecutive titles
- 1988-1989: Sito Pons (Honda)
- 1990: John Kocinski (Yamaha)
- 1997: Max Biaggi (Honda)
- 1999: Valentino Rossi (Aprilia)
- 2001-2003: Daijiro Kato/Manuel Poggiali/Dani Pedrosa
- 2005: Dani Pedrosa (Honda)

- 2009: Hiroshi Aoyama (Honda) - Final 250cc Champion

125cc Class Champions (1949-2011) / Moto3 (2012-Present)

125cc Notable Champions:

- 1969-1970, 1972: Ángel Nieto (Derbi/Kreidler) - 13 total championships across classes
- 1971, 1973-1975, 1977, 1981-1984: Ángel Nieto
- 1987: Fausto Gresini (Garelli)
- 1990: Loris Capirossi (Honda)
- 1991: Loris Capirossi (Honda)
- 1996: Haruchika Aoki (Honda)
- 1997: Valentino Rossi (Aprilia)
- 2003: Dani Pedrosa (Honda)
- 2011: Nicolas Terol (Aprilia) - Final 125cc Champion

Moto3 Champions (2012-Present):

- 2012: Sandro Cortese (KTM)
- 2013: Maverick Viñales (KTM)
- 2014: Alex Márquez (Honda)
- 2015: Danny Kent (Honda)

- 2016: Brad Binder (KTM)
- 2017: Joan Mir (Honda)
- 2018: Jorge Martín (Honda)
- 2019: Lorenzo Dalla Porta (Honda)
- 2020: Albert Arenas (KTM)
- 2021: Pedro Acosta (KTM)
- 2022: Izan Guevara (GasGas)
- 2023: Jaume Masiá (KTM)
- 2024: David Alonso (GasGas)

Moto2 Champions (2010-Present)

- 2010: Toni Elías (Moriwaki)
- 2011: Stefan Bradl (Kalex)
- 2012: Marc Márquez (Suter)
- 2013: Pol Espargaró (Kalex)
- 2014: Tito Rabat (Kalex)
- 2015: Johann Zarco (Kalex)
- 2016: Johann Zarco (Kalex)
- 2017: Franco Morbidelli (Kalex)
- 2018: Francesco Bagnaia (Kalex)
- 2019: Álex Márquez (Kalex)
- 2020: Enea Bastianini (Kalex)
- 2021: Remy Gardner (Kalex)
- 2022: Augusto Fernández (Kalex)

- 2023: Pedro Acosta (Kalex)
- 2024: Ai Ogura (Boscoscuro)

Circuit Specifications and Lap Records

Current MotoGP Calendar Circuits (2024 Season)

Circuit of the Americas (COTA) - Austin, Texas, USA

- Length: 5.513 km (3.426 mi)
- Turns: 20 (11 left, 9 right)
- First MotoGP race: 2013
- Lap record: Marc Márquez (Honda) - 2:02.135 (2014)
- Notable features: Elevation changes, Turn 1 uphill complexity
- Capacity: 120,000 spectators

Autodromo Internazionale del Mugello - Italy

- Length: 5.245 km (3.259 mi)

- Turns: 15 (6 left, 9 right)
- First Grand Prix: 1976
- Lap record: Danilo Petrucci (Ducati) - 1:45.245 (2019)
- Notable features: High-speed straights, challenging Casanova-Savelli complex
- Capacity: 112,000 spectators

Circuit de Barcelona-Catalunya - Spain

- Length: 4.675 km (2.905 mi)
- Turns: 14 (5 left, 9 right)
- First Grand Prix: 1992
- Lap record: Fabio Quartararo (Yamaha) - 1:38.853 (2021)
- Notable features: Long front straight, technical sector 3
- Capacity: 140,700 spectators

Red Bull Ring - Spielberg, Austria

- Length: 4.318 km (2.683 mi)
- Turns: 10 (3 left, 7 right)
- First MotoGP race: 1997 (returned 2016)
- Lap record: Jorge Martín (Ducati) - 1:28.596 (2023)

- Notable features: Shortest circuit, high altitude (678m)
- Capacity: 72,000 spectators

Silverstone Circuit - Great Britain

- Length: 5.891 km (3.660 mi)
- Turns: 18 (8 left, 10 right)
- First Grand Prix: 1977
- Lap record: Jorge Martín (Ducati) - 1:57.767 (2023)
- Notable features: High-speed corners, Maggotts-Becketts complex
- Capacity: 150,000 spectators

Misano World Circuit Marco Simoncelli - Italy

- Length: 4.226 km (2.626 mi)
- Turns: 16 (6 left, 10 right)
- First Grand Prix: 2007
- Lap record: Francesco Bagnaia (Ducati) - 1:31.065 (2023)
- Notable features: Technical layout, named after Marco Simoncelli
- Capacity: 60,000 spectators

Motegi Twin Ring - Japan

- Length: 4.801 km (2.983 mi)
- Turns: 14 (6 left, 8 right)
- First Grand Prix: 2000
- Lap record: Marc Márquez (Honda) - 1:43.506 (2019)
- Notable features: Honda's home circuit, unique figure-8 design
- Capacity: 50,000 spectators

Phillip Island Grand Prix Circuit - Australia

- Length: 4.448 km (2.764 mi)
- Turns: 12 (4 left, 8 right)
- First Grand Prix: 1989
- Lap record: Marc Márquez (Honda) - 1:27.899 (2017)
- Notable features: Coastal location, high-speed flowing corners
- Capacity: 100,000 spectators

Sepang International Circuit - Malaysia

- Length: 5.543 km (3.444 mi)
- Turns: 15 (8 left, 7 right)
- First Grand Prix: 1999

- Lap record: Jorge Lorenzo (Ducati) - 1:58.303 (2018)
- Notable features: Tropical climate challenges, Hermann Tilke design
- Capacity: 130,000 spectators

Losail International Circuit - Qatar

- Length: 5.380 km (3.343 mi)
- Turns: 16 (6 left, 10 right)
- First Grand Prix: 2004
- Lap record: Jorge Martín (Ducati) - 1:52.772 (2023)
- Notable features: Night race under floodlights, desert setting
- Capacity: 33,000 spectators

Historic Circuits No Longer on Calendar

Assen TT Circuit - Netherlands (Still Active)

- Length: 4.542 km (2.822 mi)
- Known as: "The Cathedral of Speed"
- First Grand Prix: 1949 (continuous since then)
- Lap record: Francesco Bagnaia (Ducati) - 1:31.504 (2023)

Spa-Francorchamps - Belgium (Last race: 1990)

- Length: 6.985 km (Original circuit)
- Notable features: Legendary Eau Rouge corner, extreme danger
- Last lap record: Wayne Gardner (Honda) - 2:34.24 (1987)

Isle of Man TT Course - United Kingdom (Last championship race: 1976)

- Length: 60.725 km (37.73 mi)
- Notable features: Public roads, most dangerous circuit ever used
- Still hosts Tourist Trophy races annually

Brands Hatch - United Kingdom (Last race: 1986)

- Length: 4.206 km (Grand Prix circuit)
- Notable features: Natural amphitheater, Paddock Hill Bend
- Hosted British Grand Prix 1977-1986

Statistical Analysis and Record Holders

Most Premier Class Championships

1. **Giacomo Agostini (Italy)** - 8 championships (1966-1975)
2. **Valentino Rossi (Italy)** - 7 championships (2001-2005, 2008-2009)
3. **Marc Márquez (Spain)** - 6 championships (2013-2014, 2016-2019)
4. **Mike Hailwood (Great Britain)** - 4 championships (1962-1965)
5. **Eddie Lawson (USA)** - 4 championships (1984, 1986, 1988-1989)
6. **Mick Doohan (Australia)** - 5 championships (1994-1998)

Most Grand Prix Victories (All Classes)

1. **Giacomo Agostini** - 122 wins (68 in 500cc, 54 in 350cc)
2. **Ángel Nieto** - 90 wins (13 in 500cc, 77 in smaller classes)
3. **Valentino Rossi** - 89 wins (89 in premier class)

4. **Mike Hailwood** - 76 wins (37 in 500cc, 39 in other classes)
5. **Mick Doohan** - 54 wins (all in 500cc)

Most Premier Class Wins

1. **Giacomo Agostini** - 68 wins
2. **Valentino Rossi** - 89 wins
3. **Marc Márquez** - 59 wins
4. **Mick Doohan** - 54 wins
5. **Casey Stoner** - 38 wins

Most Pole Positions (Premier Class)

1. **Marc Márquez** - 62 poles
2. **Valentino Rossi** - 55 poles
3. **Jorge Lorenzo** - 41 poles
4. **Casey Stoner** - 39 poles
5. **Mick Doohan** - 58 poles

Youngest and Oldest Records

Youngest Premier Class Champion:

- Marc Márquez - 20 years, 266 days (2013)

Oldest Premier Class Champion:

- Nicky Hayden - 25 years, 24 days (2006)

Youngest Race Winner:

- Marc Márquez - 20 years, 63 days (Austin 2013)

Oldest Race Winner:

- Giacomo Agostini - 33 years, 257 days (Nürburgring 1975)

Most Podium Finishes (Premier Class)

1. **Valentino Rossi** - 199 podiums
2. **Giacomo Agostini** - 159 podiums
3. **Marc Márquez** - 101 podiums
4. **Jorge Lorenzo** - 114 podiums
5. **Dani Pedrosa** - 112 podiums

Constructor Championships (Premier Class)

Most Constructors' Titles:

1. **Honda** - 25 titles (1983, 1985, 1987, 1989, 1994-1999, 2001-2003, 2006, 2011, 2013-2014, 2016-2019)

2. **Yamaha** - 17 titles (1974-1975, 1978-1980, 1984-1986, 1988, 1990-1992, 2004-2005, 2008-2010, 2012, 2015, 2021)
3. **MV Agusta** - 17 titles (1956-1973)
4. **Ducati** - 3 titles (2007, 2022-2023)
5. **Suzuki** - 3 titles (1976-1977, 1993, 2020)

Team Championships (Since 2002)

Most Team Titles:

1. **Repsol Honda Team** - 8 titles
2. **Yamaha Factory Racing** - 6 titles
3. **Ducati Lenovo Team** - 2 titles
4. **Team Suzuki Ecstar** - 1 title

Manufacturer Participation History

Premier Class Constructor Participation by Era

Foundation Era (1949-1965):

- **British Manufacturers:** Norton, AJS, Velocette, Matchless
- **Italian Manufacturers:** Gilera, MV Agusta, Moto Guzzi
- **German Manufacturers:** NSU, BMW

Japanese Invasion (1966-1990):

- **Honda:** Entered 1959, first win 1966 with Mike Hailwood
- **Yamaha:** Entered 1961, breakthrough with Phil Read
- **Suzuki:** Entered 1974, Barry Sheene championships 1976-1977
- **Kawasaki:** Limited participation, focus on smaller classes

Modern Era Manufacturers (1990-Present):

- **Honda:** Continuous participation, NSR500 to RC213V evolution
- **Yamaha:** YZR500 to M1, consistent front-runner
- **Suzuki:** GSV-R development, championship with Joan Mir 2020
- **Ducati:** 2003 entry with Desmosedici, breakthrough 2007

- **Aprilia:** Brief premier class participation 2000-2004, returned 2015
- **KTM:** Entered 2017 with RC16 project

Constructor Championship Success Rate

Most Successful Eras by Constructor:

MV Agusta (1956-1976):

- Championships: 17 consecutive (1956-1973)
- Total wins: 139 races
- Notable riders: John Surtees, Mike Hailwood, Giacomo Agostini
- Technical innovation: First use of disc brakes in racing

Honda (1983-Present):

- Championships: 25 total
- Peak era: 1994-1999 (Mick Doohan dominance)
- Modern era: 2013-2019 (Marc Márquez dominance)
- Technical contributions: Pneumatic valve systems, seamless gearbox

Yamaha (1974-Present):

- Championships: 17 total
- Golden periods: 1978-1980 (Kenny Roberts era), 2004-2005, 2008-2010 (Rossi era)
- Technical innovations: Crossplane crankshaft design, advanced chassis dynamics
- Rider development: Strong junior category programs

Ducati (2003-Present):

- Championships: 3 (2007, 2022-2023)
- Unique approach: L-twin engine configuration until 2020
- Technical philosophy: Desmodromic valve system, carbon fiber chassis
- Recent success: V4 engine development with Bagnaia championships

Failed Manufacturer Attempts

Ilmor Engineering (2005-2006):

- Three-cylinder 800cc prototype
- Partnership with various teams
- Technical challenges with competitiveness
- Withdrew after limited success

Blata (2014-2016):

- Czech manufacturer attempt
- Forward Racing partnership
- Struggled with reliability and performance
- Ceased operations due to financial constraints

WCM (1998-2000):

- British constructor with Yamaha engines
- Limited budget and resources
- Failed to score championship points
- Demonstrates challenges for privateer constructors

Current Constructor Competitiveness Analysis (2020-2024)

Tier 1 - Championship Contenders:

- **Ducati:** 3 championships, 47 wins, technological leadership
- **Yamaha:** 1 championship, 12 wins, consistent podium threat
- **Honda:** 0 championships, 8 wins, rebuilding phase

Tier 2 - Race Winners:

- **KTM:** 0 championships, 11 wins, rapid development progress
- **Suzuki:** 1 championship, 3 wins, withdrew after 2022

Tier 3 - Point Scorers:

- **Aprilia:** 0 championships, 3 wins, steady improvement trajectory

Printed in Dunstable, United Kingdom